AKADEMIE DER WISSENSCHAFTEN DER DDR

ACTA HYDROPHYSICA

Begründet von HANS ERTEL

Im Auftrage des Forschungsbereichs Geo- und Kosmoswissenschaften
herausgegeben von
S. DYCK, P. MAUERSBERGER, K. VOIGT

Band XXXI, Heft 3/4

Mit 27 Abbildungen und 7 Tabellen

AKADEMIE-VERLAG BERLIN
1987

Die ACTA HYDROPHYSICA erscheinen unter Mitwirkung von
O. Czepa, H. Hartke, G. Schellenberger

Anschriften der Herausgeber

Prof Dr S. Dyck
Technische Universität Dresden, Sektion Wasserwesen,
Bereich Hydrologie und Meteorologie
DDR-8027 Dresden, Mommsenstr 13

Prof Dr P Mauersberger
Akademie der Wissenschaften der DDR,
Institut für Geographie und Geoökologie, Bereich Hydrologie
DDR-1162 Berlin, Müggelseedamm 260

Prof Dr K. Voigt
Akademie der Wissenschaften der DDR,
Institut für Meereskunde
DDR-2530 Rostock-Warnemünde. Postschließfach 38

Redaktion

Dr K.-H Hohl
Akademie der Wissenschaften der DDR
DDR-1086 Berlin, Leipziger Str 3—4

Manuskriptsendungen werden an einen der Herausgeber erbeten. Es können Originalarbeiten in deutscher, russischer und englischer Sprache eingereicht werden Die Autoren werden gebeten, an den Anfang jeder Arbeit Zusammenfassungen in deutscher und englischer (Summary) sowie ggf in russischer Sprache (Резюме) zu stellen. Von jedem Beitrag werden 50 Sonderdrucke kostenfrei geliefert.
Die ACTA HYDROPHYSICA erscheinen mit 4 Heften pro Band im Format A5

ISBN 3-05-500262-8
ISSN 0065-1338

Erschienen im Akademie-Verlag Berlin, DDR-1086 Berlin, Leipziger Straße 3—4

Lizenznummer 202 100/440/87
Printed in the German Democratic Republic
Gesamtherstellung VEB Druckerei „Thomas Müntzer", 5820 Bad Langensalza
LSV 1465
Bestellnummer 763 680 7 (2050/31/3—4)
03600

AKADEMIE DER WISSENSCHAFTEN DER DDR

ACTA HYDROPHYSICA

Begründet von HANS ERTEL

Im Auftrage des Forschungsbereichs Geo- und Kosmoswissenschaften
herausgegeben von

S. DYCK, P. MAUERSBERGER, K. VOIGT

Band XXXI

AKADEMIE-VERLAG BERLIN
1987

Inhalt des Bandes XXXI

ISSN 0065—1338

Inhaltsverzeichnis

Acta hydrophys., Berlin **31** (1987) 3/4, S. 125—141

EGMO — System Approach and Subroutines for River Basin Modeling

By A. Becker and B. Pfützner[1])

Summary River basin models are required for different purposes in operational hydrology and water resources systems planning, design and management. A model system approach has been developed which allows to construct different versions of river basin models by combining a number of well-established simple submodels for subprocesses of the hydrological cycle. After a brief description of this system approach EGMO, and of some important submodels, selected results of application of the real-time discharge forecasting version EGOP of the model system EGMO are presented.

Zusammenfassung Flußeinzugsgebietsmodelle werden für verschiedene Zwecke in der operativen Hydrologie, bei der Untersuchung, Planung und Bewirtschaftung von Wasserressourcensystemen u. ä. benötigt. Es wurde ein Modellsystemkonzept entwickelt, daß es erlaubt, durch geeignete Kombinationen von erprobten, einfachen Teilmodellen für die beteiligten hydrologischen Prozesse verschiedenen Formen von Flußeinzugsgebietsmodellen zu erstellen Nach einer kurzen Beschreibung dieses Systemkonzepts EGMO und einiger der wichtigsten Teilmodelle werden einige Ergebnisse der Anwendung der Modellversion EGOP des Modellsystems EGMO vorgestellt, mit der Echtzeitvorhersagen des Durchflusses gegeben werden können.

Резюме Для различных целей оперативной гидрологии, а также в области системы водохозяйственного использования, планирования и управления необходимы различные формы моделей речных бассейнов. Для этого была разработана концепция системы моделей, которая позволяет составлять эти формы с помощью подходящей комбинации круговорота воды Дано краткое описание этой концепции системы ЭГМО и некоторых важнейших частичных моделей В заключение представлены некоторые результаты применения варианта модели краткосрочного прогнозирования протоков ЭГОП системы моделей ЭГМО

[1]) Dr sc. techn Alfred Becker, Dipl.-Hydr Bernd Pfützner, Institut für Wasserwirtschaft, Schnellerstraße 140, Berlin, DDR-1190

1. EGMO — a system approach for river basin modeling

River basin models which are capable to simulate the water cycle and the hydrological regime in river basins are required for the solution of different tasks in research and practice, in particular for

- the investigation of long-term variations in water storage, discharge and evaporation under the given and under changing climate or land use conditions,
- long-term planning of water resources use,
- real-time hydrological forecasting.

In consideration of these requirements the generalized system approach EGMO has been developed which enables the model user to construct different model forms by an appropriate combining of available generalized submodels (subroutines), according to the specific river basin conditions, requirements, constraints etc.

Most important existing and applied forms of the model system EGMO are listed and briefly characterized in table 1 ([8], [13]).

The first version of the river basin model EGMO was presented already in 1971 ([2]). This model was lumped. It included, however, for important sub-processes (e.g. soil moisture storage) the consideration of a "statistical" distribution of some model parameters and state variables within the catchment, according to the infiltration approach introduced in the Stanford model ([10]).

In the next step of development a model element was introduced which considers dynamically varying water saturated sub-areas ([3], [4]). These sub-areas which increase with increasing soil moisture are found to be the main contributors to direct overland flow, i.e. to flood flow, in particular during heavy or long rainfall and snowmelt periods ([11], [14], [16], [5]).

Based on the experience received in the development and application of these

TABLE 1
Specific model forms of EGMO developed for different purposes and tested in different river basins

Model form	Time increment	Purpose and region of application
EGMOF	≤ 1 d	longterm simulation of water storage, discharge and evapotranspiration in flatland river basins
EGFOP	≤ 30 d	real-time forecasting of river discharge in flatlands
EGOP	1 h to 6 h	as EGFOP, but in mountaineous basins
EGMOD	5 d, 10 d, 1 m	as EGMOF, but with greater time increments
EHOF	3 h to 1 d	computation of design floads (primarily for flatlands)

TABLE 2
Subareas in a river basin with significantly different hydrological regimes

Subarea type	Symbol	Evapotranspiration	Direct runoff generation
• Open water surfaces	*AW*	potential	equal to total precipitation (100%)
• Impervious (rocks, paved etc.)	*AIMP*	only during rainfall, i.e. near to zero	near to total precipitation (near to 100%)
Flat areas with:			
• deep groundwater	*AF*	soil moisture dependent	near to zero
• shallow groundwater	*AN*	generally potential	from saturated subareas *AS* (which vary moisture dependent) near to total precipitation
Sloped areas with:			
• deep permeable underground	*AF*	as *AF* (above)	as *AF* (above), infiltration excess in case of heavy rainfall or frozen ground possible
• shallow soils, highly permeable subsoils	*AH*	soil moisture dependent	interflow (quick return) and infiltration excess or *AS*-overland flow (as from *AN*)

basic model versions the system approach (called EGMO) was developed in which the following general requirements were fulfilled:

— Four runoff components and the corresponding "runoff levels" can be considered:
  - overland flow,
  - interflow (quick return lateral subsurface flow through permeable soil horizons or zones),
  - shortterm base flow,
  - longterm base flow.
— A zoning approach can be applied by which subareas as listed in Table 2 which behave typically different in evapotranspiration and runoff regime can be simulated separately.
— Computation time increments from 1 hour to 1 month can be used.
— Physically based simplified and conceptual submodels for all subsystems and subprocesses were preferred which allow to derive model parameters from related basin characteristics or for transferring them from neighbouring similar river basins (when in the interesting basin no or only insufficient hydrological observation records exist).

Rather simple but well established and powerful subsystem models have been developed and successfully tested with different computation time increments for a number of catchments with different conditions:

(a) Interception submodel,
(b) soil moisture accounting procedure with special reference to tension water content and to its areal distribution according to field capacity as a distributed parameter within the considered subareas (particularily *AF*),
(c) submodels for soil gravity water and groundwater storage and flow in the subareas *AF*,
(d) analoguous submodels for the subareas *AN*,
(e) submodels for determining during periods of heavy or longperiod rainfall and/or snowmelt saturated subareas which produce direct overland flow,
(f) submodels for overland and channel flow dynamics.

For the interception submodel (a) a simple storage element of about 5 mm storage capacity, filled by precipitation and exhausted by evapotranspiration, has been found sufficient.

The soil moisture accounting model as mentioned under (b) is extremely important for the simulation of runoff formation and also of evapotranspiration. Therefore it will be described in the following. All other submodels listed above, except that developed for subareas *AN* with shallow groundwater (mentioned under (d) and described in another paper [9]), will be briefly characterized in the following in their simplest form, with special concern to the real-time forecasting version EGOP of the model system EGMO.

2. Soil moisture recharge on a basin-wide scale

The principle structure of the operational forecasting version of the river basin model EGOP is represented in Fig. 1. Symbols see list of symbols pp. 139—141.

Model input *PO* is the difference between precipitation (real corrected data *P(I)*) and potential evapotranspiration (*EP(I)*) of the time step *DT* under consideration. *PO* is positive for runoff formation intervals, otherwise it is negative (exhaustion interval).

That part of *PO* that falls in the sub-areas *AW* and *AIMP* is taken unreduced as first direct input component into the channel system:

$$RO_D = PO \cdot (AW + AIMP) \tag{1}$$

In the remaining part ($1 - AW - AIMP$) of the total catchment area *AE* is initially assumed that $PO(= PB)$ will totally infiltrate. The loss computation, i.e. the computation of losses from the infiltrating water (*PB*) caused by the capillary water recharge $\mathrm{d}WS/\mathrm{d}t$ in the unsaturated and rooted zone of the soil, is based on the following consideration:

There are generally holes, root channels and other "preferred pathes for infiltration" in the soil, even on nearly homogenous plots (see e.g. [15]). Therefore,

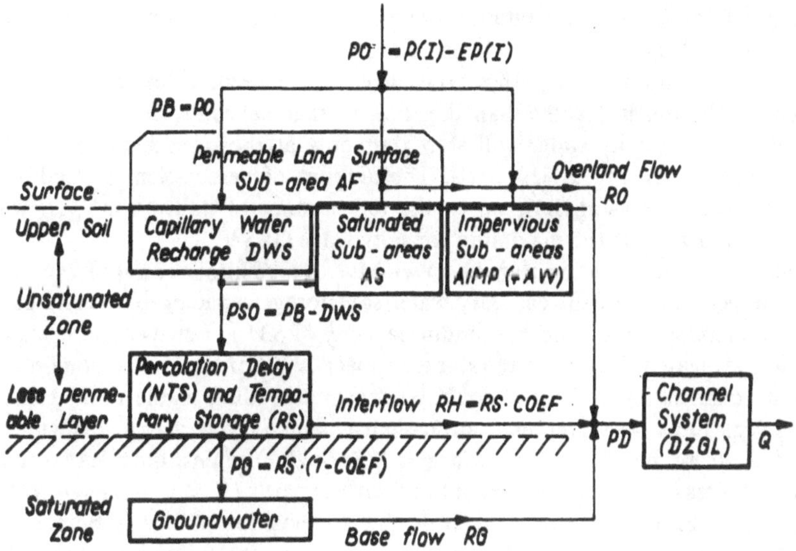

Fig. 1. Structure scheme of the operational forecasting model EGOP with indication of the flow components considered

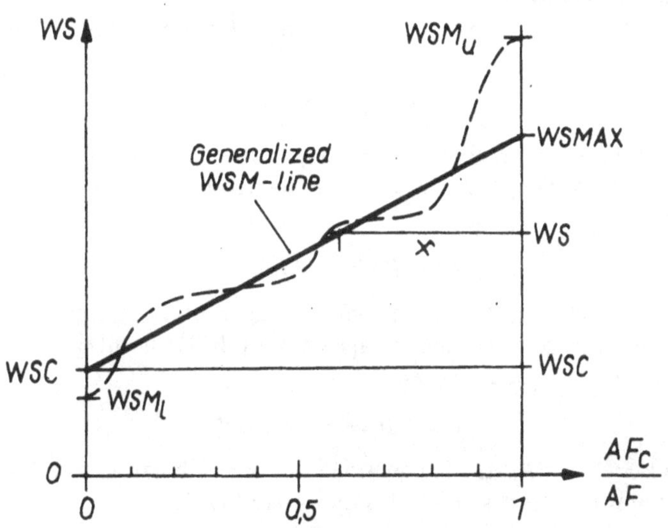

Fig. 2. Real (dashed line) and generalized distribution (solid line) of the soil storage capacity for capillary water (*WSM*) within a river basin

during rainfall events precipitation water can infiltrate at some places immediately into deeper horizons. After filling of the current moisture deficiency around these "infiltration pathes" (i.e. when WS exceeds a lower limit WSM_1 which can be much smaller than the total storage capacity of the soil profile) percolation of gravity water will start, but only at those places where WSM_1 was exceeded (see Fig. 2, left part). The amount of percolation is steadily increasing with increasing WS, up to the full amount of infiltration (PB) when WS will approximate the maximum capacity value (WSM_u).

Considering now a larger river basin we have to take into account in any case a much larger variety of the capillary water soil storage capacity (between a basin-wide minimum WSM_1 and maximum capacity WSM_u). Let us assume that in a well investigated river basin the storage capacites WSM are known at all places, then a "distribution function" could be drawn according to the dashed line in Fig. 2. This function is a representation of all available WSM-data in an increasing order (i.e. from the smallest to the largest WSM-value) against the releated cumulative sub-areas AF_C of the permeable land surface part ($AF = 1 - AW - AIMP$) of the total basin area (not considering the real position of each sub-area). For simplification the sub-area AN is not considered in the following. Generally a nonlinear function will be received (as e.g. the dashed line in Fig. 2).
This function may be approximated by a linear relation (solid line in Fig. 2) which defines the generalized parameters WSC and $WSMAX$ of the river basin. According to this relation it can be said:

(1) If WS is smaller than WSC, then the full area AF is contributing to soil moisture recharge:

$$\mathrm{d}WS/\mathrm{d}t = (PB/DT) \tag{2}$$

where (PB/DT) is the areal average of the infiltration rate during the time step $DT = t - t_1$. By integration over DT the total recharge $DWS = WS - WS_1$ is received:

$$DWS = PB \tag{3}$$

(2) If WS is greater than WSC, then only that part x of AF can contribute to soil moisture recharge where the storage capacity WSM is still greater than WS (right upper triangle in Fig. 2):

$$\mathrm{d}WS/\mathrm{d}t = x \cdot (PB/DT) \tag{4}$$

where $x = 1 - AF_c/AF$. In equation (2) is $x = 1$ because of $AF_c/AF = 0$. x can be expressed in terms of WS as follows (see Fig. 2)

$$\frac{x}{1} = \frac{(WSMAX - WS)}{(WSMAX - WSC)} \,. \tag{5}$$

Inserting this equation into (4) we receive

$$\frac{dWS}{dt} = \frac{PB}{DT} \cdot \frac{(WSMAX - WS)}{(WSMAX - WSC)} \tag{6}$$

$D = (WSMAX - WS)$ represents a moisture deficiency for which equation (6) reads

$$-\frac{dD}{dt} = \frac{PB}{DT} \cdot \frac{D}{(WSMAX - WSC)} . \tag{7}$$

Assuming PB = const during a time step $DT = t - t_1$ (and a decrease of D from D_1 to D) a solution can be derived ([1]):

$$\left.\frac{dD}{D}\right|_{D_1}^{D} = -\frac{PB}{DT \cdot (WSMAX - WSC)} \left. dt \right|_{t_1}^{t}$$

or

$$\ln D - \ln D_1 = -PB/(WSMAX - WSC) \tag{8}$$

or

$$D = D_1 \cdot \exp\left(-PB/(WSMAX - WSC)\right) . \tag{9}$$

Now the total soil capillary water recharge DWS in the area AF during the time step DT can be determined:

$$DWS = WS - WS_1 = D_1 - D .$$

Inserting here equation (9) and $D_1 = (WSMAX - WS_1)$ gives:

$$DWS = (WSMAX - WS_1) \cdot [(1 - \exp) \cdot (-PB)/(WSMAX - WSC)] . \tag{10}$$

It has been found in a number of test computations that the interception capacity of about 2 to 5 mm can be included in the soil moisture capacity parameters $WSMAX$ and WSC.

From DWS and PB the areal average of soil percolation or gravity water discharge PSO in the area AF during the reference time interval DT can be determined:

$$PSO = PB - DWS$$

PSO is input to the subsequent flow systems (interflow RH, base flow RG; see Fig. 1).

3. Moisture dependent saturated sub-areas and total overland flow

There are different possibilities to simulate infiltration excess in a river basin. According to results in a number of experimental basins ([11], [14], [16], [5]) infiltration excess is limited in many cases to small sub-areas in a river basin which are increasing with continuing rain and increasing soil moisture. Therefore the simple conception of moisture dependent sub-areas which was already introduced in 1975 ([3], [4]) is used in the EGMO-System. This conception assumes a linear relation between the saturated sub-area *AS* and a soil moisture index, as e.g. the current soil moisture storage (*WS*) or gravity water discharge (*PSO*). In the operational forecasting model EGOP the following relation is applied:

$$AS = PSO/FMAX \tag{12}$$

where *FMAX* is a parameter (dimension: mm/*DT* · h). It is necessary to additionally define *ASMAX* as maximum possible saturated sub-area *AS* in the river basin (if *AS* in eq. (12) is greater than *ASMAX* then *AS* = *ASMAX*).
With given *AS* the saturated area runoff RO_S as main component of overland flow can be determined (see Fig. 1):

$$RO_S = PO \cdot AS\,. \tag{13}$$

By combining equations (1) and (13) the total overland flow is received as:

$$RO = RO_D + RO_S = PO \cdot (AW + AIMP + AS) \tag{14}$$

4. Delayed flow components

That part of the catchment area which is not included in equation (14) is assumed to produce no overland flow ($AF = 1 - AW - AIMP - AS$). Precipitation water infiltrates here completely and produces interflow (*RH*) or groundwater recharge (*PG*), An average time delay *TS* needed for water percolation through the upper soil horizons is introduced by a simple delay element *NTS* (= *TS*/*DT*):

$$PS\,(NTS + 1) = PSO \cdot (1 - AW - AIMP - AS) \tag{15}$$

The delayed percolation *PS*(1) represents the input to the interflow system (soil gravity water storage above less pervious soil horizons; see Fig. 1). This system is represented by a single linear reservoir which has been proved as a sufficient approximation in a number of case studies (e.g. [7]). The outflow *RS* of such a reservoir at the end of a time step *DT* (with an input *PS*(1) of constant intensity during *DT*) can be computed from the outflow *RS*1 at the beginning of *DT* as follows ([12], chapter 4):

$$RS = CSE \cdot RS1 + (1 - CSE) \cdot PS(1) \tag{16}$$

where $CSE = 1/EXP\,(DT/CS)$
CS — storage constant of the soil gravity water reservoir.
With $PS(1) = 0$ equation (16) represents a recession formula ($RS = RS1 \cdot CSE$). From the sloped part AH of subarea AF RS runs of laterally as interflow (see Fig. 1):

$$RH = RS \cdot AH \tag{17}$$

where AH as a part of AF is dimensionless ($0 \leqq AH < 1$).
The remaining part of RS becomes groundwater recharge:

$$PG = RS - RH = RS \cdot (1 - AH)\,. \tag{18}$$

Similar to the soil gravity water storage system (or interflow system) the groundwater system can be represented in a first order approximation by a single linear reservoir. Therefore base flow RG can be computed by the same simple relation (see equation (16)):

$$RG = CGE \cdot RG1 + (1 - CGE) \cdot PG \tag{19}$$

where $CGE = 1/EXP\,(DT/CG)$
CG — storage constant of the groundwater reservoir.

5. Computation of river basin discharge

The total inflow into the channel system is composed of the three flow components defined above:

$$PD(M) = (RG + RH - RO) \cdot AET \tag{20}$$

$AET = AE/3.6 \cdot DT)$
AE — river basin area (km^2).
This inflow has to pass the channel system (see Fig. 1), which in many cases, at least in mountainous or hilly river basins with sufficient slope to the closing profile, behaves approximately as a linear system. Thus, the channel system operation can be represented by the convolution equation, e.g. in the discrete form:

$$QC(M) = \sum_{I=1}^{N} H(I) \cdot PD(M - I + 1) \tag{21}$$

where $QC(M)$ — computed basin discharge in discrete time steps DT,
$H(I)$ — discrete ordinates of the DT-impulse response function (unit hydrograph) of the channel system ($1 \leqq I \leqq N$).

The ordinates $H(I)$ can be derived in a simple way from basin outflow hydrographs observed during summer storm events of short duration and high approximately uniform rainfall intensity within the river basin, especially in periods when the soil moisture storage WS is so small that no remarkable interflow occurs. In such cases the basin outflow hydrograph represents mainly overland flow RO which is only transformed by the channel system.

It is well-known that river basin responses are highly nonlinear. In the catchment model EGOP this nonlinerarity is reflected by different percentages of the three considered flow components RG, RH and RO in $PD(M)$ from event to event. These different percentages are caused by the different formation conditions and different frequency characteristics of each component:

- RO is directly dependent on PO (only partially on PSO, i.e. on WS). It occurs only during rainfall intervals and can therefore change from one time step to the other from zero to a large value and back to zero (high frequent signal).
- RH is dependent more on WS and only indirectly dependent on PO, but additionally on $RS1$ and AH. It is delayed against RO by NTS time steps and transformed through a linear reservior (soil system) with the storage constant CS (medium frequent signal).
- RG is additionally delayed against RH and extremely flattened through a second linear reservoir (the groundwater system) with a much larger storage constant CG.

Therefore the large variety of observed outflow hydrographs $Q(M)$ is quite obvious.

6. Updating of discharge forecasts

Convolution equations (as eq. (21)) and difference equations (ARMA-typ models) have been proved to be completely equivalent forms for describing the input-output transformation in a linear system. The difference equation reads in a generalized form (with $QC(M)$ and $PD(M)$ as in eq. (21)):

$$QC(M) = \sum_{J=1}^{JMAX} Q(M-J) \cdot A(J) + \sum_{I=IT}^{NR} PD(M-I) \cdot B(I) \qquad (22)$$

In a recently published paper ([6]) simple relations have been derived for determining from given unit- hydrograph ordinates $H(I)$:

(1) The number IT of initial zero-ordinates ($H(I) = 0$), i.e. the so-called "dead-time" at the beginning of the channel system response function $H(I)$. Most often this dead-time is zero ($IT = 0$),

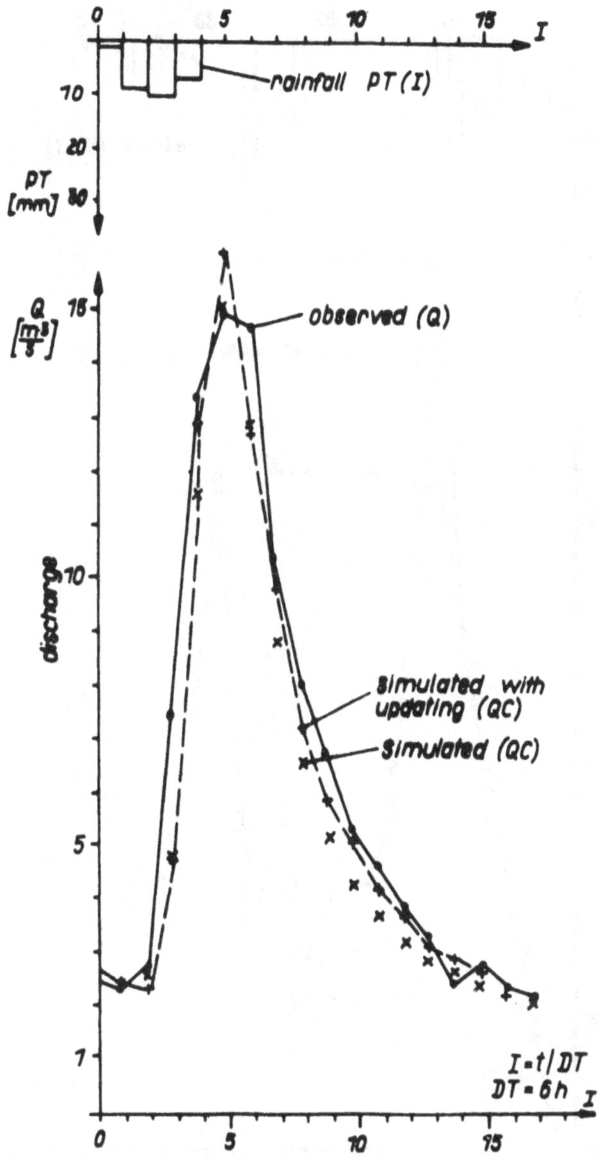

Fig. 3 Observed and simulated flood discharges of a test catchment in the upper Spree basin (21 8 –24 8 1977)

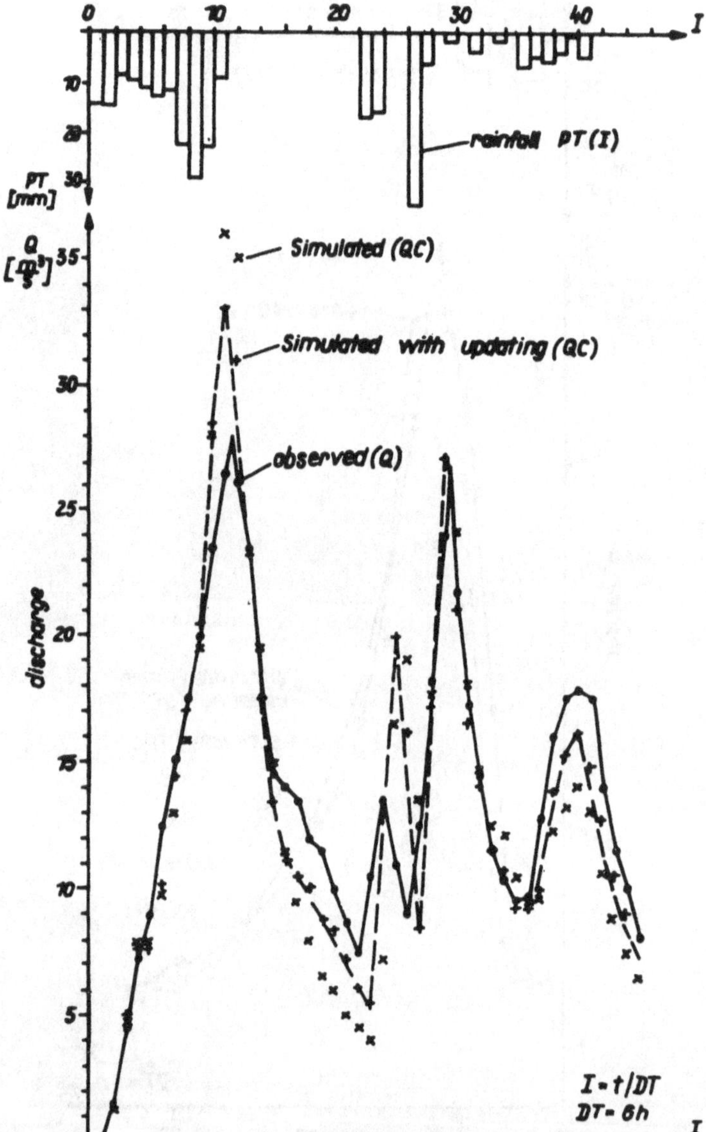

Fig. 4 Observed and simulated flood discharges of a test catchment in the upper Spree basin (18 7 – 31 7 1981)

(2) the order NR of the difference equation (number of the inflection point ordinate of the unit-hydrograph recession),
(3) the coefficients $A(J)$ and $B(I)$ of the difference equation.

Since in equation (22) the last observed discharges $Q(M - J)$ are included when $QC(M)$ is forecasted one step ahead, an automatic updating is achieved which is very valuable and efficient in real-time forecasting.
To demonstrate the efficiency of the model EGOP and of the updating procedure according to equation (22) two examples of 6 h-forecast for a tributary sub-basin of the upper Spree (catchment area $AF = 135$ km^2) are represented in Fig. 3 and 4. In Fig. 3 observed and forecasted discharge hydrographs of a normal rainfall flood in August 1977 are presented. Fig. 4 refers to a sequence of flood waves resulting from a sequence of heavy summer rainstorms in July 1981. Some additional remarks should be given on the application of equation (22):

— In the given form eq. (22) is valid for 1-step-ahead-forecasts (e.q. 6 hour forecasts as in Fig. 3 and 4) for which in cases when $IT = 0$ 6 h-forecasts of rainfall and snowmelt are needed.
— If discharge forecasts are wanted for more time steps ahead then the river basin discharges $QC(M)$ previously forecasted have to be taken as independent variables $Q(M)$ on the right side of eq. (22) (of course only for those future time steps where $Q(M)$ is still unknown. For these time steps rainfall or snowmelt forecasts must also be given or assumed as zero).
— Accordingly eq. (22) can be principally applied with computed basin discharges for simulation runs. This was done for getting the simulation results in Fig. 3 and 4 (open points).

Thus it is clear that equation (22) can be used for different purposes with simple modifications in the input data.

7. Consideration of evapotranspiration

For a more general application of the EGOP model it is necessary to include an evapotranspiration sub-procedure. This is important even during flood periods e.g. when heavy rainstorms occur as a sequence of rainfall periods with intervals of one to some days between eachother (as in Fig. 4). Between the rainfall intervals soil moisture is exhausted by evapotranspiration and free storage volume is originated which has to be refilled before direct runoff can occur again.

The evapotranspiration procedure is very similar to the moisture recharge procedure described in chapter 2. Two cases are considered in dependence of a soil moisture limit $WSGR$ (similar to WCS):

— If $WS > WSGR$ then in the area $(1 - AIMP)$ the potential evapotranspiration EP (or the difference $PB = P(I) - EP(I)$, respectively) is fully exhausted:

$$DWS = PB\,. \tag{23}$$

A difference between potential (EP) and real evapotranspiration (ER) is only given in the sub-area $AIMP$ (where $ER = 0$). Thus for the river basin as a whole is received:

$$ED = -PB \cdot AIMP\,. \tag{24}$$

– If $WS \leqq WSGR$ then the following relation is applied:

$$DWS = (EXP(PB/WSGR) - 1) \cdot WS$$

and

$$ED = -PB \cdot AIMP - (\mathrm{PB} - DWS) \cdot (1 - AIMP)\,. \tag{26}$$

In both cases the soil moisture reduction is simply $WS + DWS$ and the real river basin evapotranspiration reads:

$$ER(I) = EP(I) - ED \tag{27}$$

This approach has been proved as succesful in a number of longterm simulations with the model version EGMOF.

8. Concluding remark

The following particularities and advantages of the model system EGMO should be pointed out:

- The separation of the total process into a sequence of clearly defined component processes.
- The use of simple but physically-based sub-models for the component processes.
- The capability of the model to generate a variety of different forms of discharge hydrographs in dependence on different percentages of the three considered flow components (base flow, interflow, overland flow).
- The consideration of saturated moisture dependent dynamically varying areas in the river basin which originate during longer or heavy rainfall and snowmelt events direct overland flow and by the way extreme flood flow.
- The simple possibility of up-dating discharge forecasts by using difference equations as final component model for basin discharge computations.
- The possibility to overtake submodels and accordingly their parameters from one model version to the other.

No final state has been reached in the development and application of the model system EGMO. However, the results received are very encouraging so that additional investigations and improvements of the model system are considered as useful, in particular on a meso or macro scale basis.

List of Symbols

1. Parameters

1.1. Area parameters

$AE\,(\mathrm{km}^2)$	—total catchment area
$AET = AE/3.6 \cdot DT)$	
$AIMP$	—imperious sub-areas draining directly into the channel system
$AF = (1 - AIMP - AW)$	—permeable part of the area AE
$ASMAX$	—maximum possible saturated sub-area in a river catchment
AW	—open water surfaces
AH	— sloped subpart of subarea AF, related to $AF (0 \leqq AH < 1)$

1.2. Storage capacities and infiltration parameters

$FMAX\left(\frac{\mathrm{mm}}{DT \cdot \mathrm{h}}\right)$	— parameter related to the area [illegible] of saturated hydraulic conductivity
WSC (mm)	—generalized minimum value of the soil moisture storage capacity for capillary water (within area AF)
$WSGR$ (mm)	—according to WSC, eventually smaller
$WSMAX$ (mm)	—generalized maximum value of the soil moisture storage capacity for capillary water (within area AF)

1.3. Parameters defining flow dynamics

$A1, A2$	—parameters of the difference equation (ARMA-type model) describing the behaviour of the channel system
IT	—number of the first ordinate of $B(J)$ which exceeds zero ($IT > 0$)
NR	—number of the last ordinate of $B(J)$ which exceeds zero
NTS	—number of time steps DT needed in the average for water percolation through the upper soil horizon
CG(d)	—storage constant of the baseflow system (groundwater system) considered to behave as linear reservoir
$CGE = 1/EXP(DT/CG)$	
CS(d)	—storage constant of the soil gravity water system (see CG)

1.4. Control parameters

DT(h) —duration of one computation time step *I*
IMAX —total number of time steps considered
UPDATE —parameter deciding on the use of the updating feature of the channel system model (*UPDATE* = 1) or on its excluding (*UPDATE* = 0)

2. Input and output data

2.1. Precipitation (mm/*DT* · h)

PT(*I*) —rainfall and snowmelt amount during a computation time step *I* (corrected areal mean)

2.2. Evapotranspiration (mm/*DT* · h)

EP(*I*) —potential evapotranspiration of time step *I*
ER(*I*) —real evapotranspiration of time step *I*
ED = *EP*(*I*) − *ER*(*I*) — so called evapotranspiration deficiency

2.3. River basin discharge components (m^3/s)

Q(*I*) —observed total discharge
QC(*I*) —simulated toal discharge
QG(*I*) —base flow component
QH(*I*) —total subsurface flow
QJ(*I*) —*QH*(*I*) plus overland flow from area *AIMP*

3. Internal variables

3.1. Internal sub-system inputs and outputs (mm/*DT* · h)

PB — water supply to the soil surface (= *PO*)
PD(*I*) — total input of the channel system (= (*RG* + *RH* + *RO*) *AET*)
PG — groundwater recharge
PI — *PT*(*I*) − *EP*(*I*)
PO = *PB*
PSO — soil gravity water percolation
PS(*J*) — soil gravity water percolation delayed by *NTS*
RG — baseflow component (of time step *I*)
RH — interflow component (of time step *I*)

RS	— total soil discharge (of time step *I*)
RO	— overland flow component (of time step *I*)

3.2. State variables

WS (mm)	— running capillary water storage of the soil
DWS (mm)	— changes of *WS* during a computation time step *I*
AS	— saturated sub-area of the area *AF* (*PSO* — dependent)

References

[1] BECKER, A.: Threshold considerations and their general importance for hydrologic systems investigation. — Fort Collins (USA): Proc. Int. Hydrol. Symp., 1967, Vol. 1, Paper 13.

[2] BECKER, A.: Applied principles of catchment simulation. — Warshaw: Proc. Int. Symp. Math. Models in Hydrology, 1971, Vol. 2, Part 2, Paper 5/12. In: IAHS-Publ. No. 102.

[3] BECKER, A.: The integrated hydrological catchment model EGMO. — Bratislava: Int. Symp. and Workshops Appl. Math. Models in Hydrology and Water Res. Syst. In: Hydrol. Sci. Bull. **21** (1975) 1.

[4] BECKER, A.: Erfassung der feuchteabhängigen Variation der Abflußbildungs- und Verdunstungsflächen in hydrologischen Einzugsgebietsmodellen. — Acta hydrophys. **20** (1976) 4, 181—211.

[5] BECKER, A.; BORCHARDT, D.: Consequences for flow simulation from results in experimental basins. — Helsinki: Contrib. Int. Symp. Infl. Man on Hydrolog. Regime, 1980.

[6] BECKER, A.: Differenzengleichung und Faltung für operative Abflußvorhersagen. Wechselbeziehungen, Vor und Nachteile. — Wasserwirtsch. Wassertechn. **32** (1982) 9, 306—308.

[7] BECKER, A.: Neue Ergebnisse und Prinzipien von Durchflußrückgangsanalysen. — Wasserwirtsch. Wassertechn. **33** (1983) 12.

[8] BECKER, A.: Grundlagen, Einzugsgebietsmodelle und Arbeitstechniken zur Berechnung von Durchflußreihen aus meteorologischen Größen. — Mitt. Inst. Wasserw. Berlin **(1984)** 46.

[9] BECKER, A.; PFÜTZNER, B.: Identification and modelling of river flow reductions caused by evapotranspiration losses from shallow groundwater areas. — Budapest: 2nd Sc. Assembly IAHS, Symp. S 2, July 1986. In: IAHS-Publ. No. 156.

[10] CRAWFORD, N. H.; LINSLEY, R. K.: Digital simulation in hydrology: Stanford watershed model IV. — Stanford: University, Dept. of Civil. Eng., 1966, Techn. Report No. 39.

[11] DUNNE, T.; MOORE, T. R.; TAYLOR, C. N.: Recognition and prediction of runoff producing zones in humid regions. — Hydr. Sci. Bull. **20** (1975), 305—327.

[12] DYCK, S.; et al.: Angewandte Hydrologie. Teil 2. — Berlin: Verlag für Bauwesen, 1980.

[13] GLOS, E.: Bestimmung von Bemessungshochwasserganglinien aus Bemessungsniederschlägen für kleine Einzugsgebiete im Flachland. — Mitt. Inst. Wasserw. Berlin **(1984)** 46.

[14] GUTKNECHT, D.: Methoden der hydrologischen Kurzfristvorhersage. — Wiener Mitt. Wasser, Abwasser, Gewässer (TU Wien) **26** (1978).

[15] KIRKBY, M. J.; et al.: Hillslope hydrology. — New York: Wiley Interscience, 1978.

[16] PESCHKE, G.: Sättigungsflächenmodelle. — In: DYCK, S.; et al.: Angewandte Hydrologie. Teil 2, Kap. 9.7. — Berlin: Verlag für Bauwesen, 1980.

Acta hydrophys., Berlin 31 (1987) 3/4, S. 143—151

Investigations on Long-Term Changes of the Seston Content in a Shallow Eutrophic Lake and their Relations to Meteorological Conditions

By H. BEHRENDT, M. OLBERG and R. STELLMACHER[1])

Summary: Long-term changes of the buoyant part of seston content of the shallow eutrophic lake Müggelsee were analysed by statistical methods. Spectral feature of this parameter was compared with series of several meteorological factors. The application of the one channel autoregressive analysis to the buoyant part of seston showed peaks at periods of about 2.2, 5 and 24 years. The spectra of the climatological series signaled the existence of periods 2.2 and 5 years too. Multichannel autoregressive spectral analysis was applied to estimate the existence of relations between the main periodicities of meteorological parameters and buoyant part of seston content. A good correlation appeared, but a phase lag is visible. This phenomenon might be shown distinctly by means of numerical filtering. At present it is not possible to explain the phase lag. Causes were discussed in connection with ecosystem development.

Zusammenfassung: Mit statistischen Methoden werden Langzeitveränderungen des auftreibenden Sestonanteils in dem flachen eutrophen Müggelsee untersucht. Das spektrale Verhalten dieses Zustandsparameters wird den Spektren von verschiedenen meteorologischen Faktoren gegenübergestellt. Auf der Basis der einkanaligen autoregressiven Spektralanalyse werden für den auftreibenden Anteil des Sestons Perioden von ungefähr 2,2, 5 und 24 Jahren nachgewiesen. Die Spektren der klimatologischen Zeitreihen zeigen ebenfalls die Existenz von Perioden von 2,2 und 5 Jahren. Die mehrkanalige autoregressive Spektralanalyse wird zur Bestimmung von Beziehungen zwischen den Periodizitäten der meteorologischen Parameter und dem auftreibenden Sestonanteil genutzt. Neben guten Korrelationsbeziehungen sind Phasenverschiebungen zwischen den jweiligen Parametern erkennbar, die durch numerische Filtration der Zeitreihen deutlich gezeigt werden können. Die Ursachen für diese Phasenverschiebungen können gegenwärtig noch nicht vollständig geklärt werden.

[1]) Dr. rer. nat. HORST BEHRENDT und Dr. sc. nat. RITA STELLMACHER, Akademie der Wissenschaften der DDR, Institut für Geographie und Geoökologie, Bereich Hydrologie, Müggelseedamm 260, Berlin, DDR-1162. Dr. sc. nat. MANFRED OLBERG, Humboldt-Universität zu Berlin, Sektion Physik, WB Meteorologie und Geophysik, Müggelseedamm 256, Berlin, DDR-1162.

1. Introduction

As the time period of observations on parameters of a lake ecosystem is increasing, the studies of intraseasonal changes and long-term variations of such parameters are becoming more and more important (LUND [6]; LUND [7]; RUSSEL [14]; AHL [1]; CHELTON et al. [4]; ROBINSON [13]; BROOKS et al. [3]).

Long-term variations of ecosystem parameters might be caused not only by anthropogenic effects but also by meteorological factors. The object of this paper is to compare the spectral features of several meteorological time series with series of seston content in a shallow eutrophic lake and to study the connections between these parameters.

The water body in question is the lake Müggelsee, the greatest lake within Berlin, the Capital of the GDR. The lake has an area of 7,2 km^2 and a mean depth of 4,9 m.

Our investigation was based on the fact that violent changes were observed in the seasonal (May—October) and annual averages of measured phytoplankton biovolume. The gratest fluctuations were found for the biovolume of blue-green algae. In the time period between 1976 and 1984 the seasonal averages of blue-green algae biovolume varied between 2.5 and 20.7 mm^3/l. The seasonal average was 9.1 mm^3/l for all 8 years (HOEG [5]; HOEG, in prep.).

The analysis of all observed lake parameter series showed that such a change in the biovolume of blue-green algae cannot be caused by human influence in terms of changes in the nutrient load of lake and particular material. Therefore, our conclusion was that these changes are due to different interannual meteorological and hydrological conditions.

The analysis of time series of blue-green algae biovolume resulted in a strong correlation between the seasonal mean of biovolume of blue-green algae and the value at the same locality in the beginning of the season (May). This probably means that the biovolume of blue-green could be determined by meteorological and hydrological conditions in winter and spring and so the state of lake in the summer, too.

Let the mean light intensity in a well mixed shallow lake be

$$\bar{I} = \frac{I_0}{\varepsilon z} \cdot (1 - e^{-\varepsilon z})$$

where z is the mean depth, ε the coefficient of extinction and I_0 the photosynthetical part of global radiation at the water surface. RAMBERG [12] defined this expression as the effective light climate.

Our investigations showed a strong nonlinear correlation between this effective light climate in March and April and the value of biovolume of the blue-green

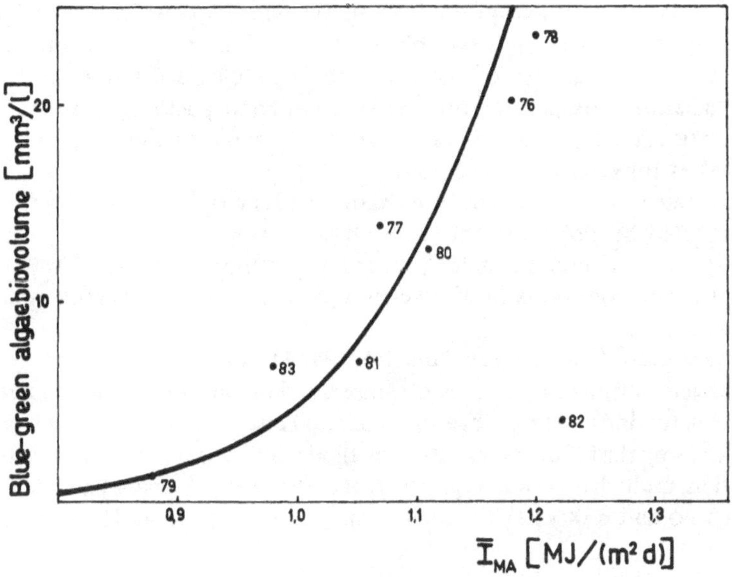

Fig. 1 Relation between the mean light climate from March and April ($\bar{I}_{MA}$) and mean blue-green algae biovolume from May for different years in the Lake Müggelsee

algae in May for the last 8 years (see Fig. 1). For confirming these results it is necessary to repeat the same tests for a longer time period.

2. Data and methods for their statistical analysis

The basis for the analysis was given by diurnal measurements of total seston wet volume in the lake, collected by the waterworks Berlin-Friedrichshagen during the past 50 years. These measurements of seston content demand the separated determination of its sinking and buoyant parts. In this way, two longtime series are available which are quite different in their statistical structure. The series of sinking suspended matter shows little interannual fluctuations but a distinct non-linear trend. This tendency can be caused by eutrophication due to human influence. Contrary to it the series of the buoyant part of seston wet volume shows violent annual changes but no tendency. In the main point the buoyant part consists of several blue-green algae, especially *Microcystis aeroginosa* and *Aphanizomenon flos-aquqe*.

In the following our investigations are concentrated on the time series of buoyant suspended matter for the time period from 1936 to 1983, which characterizes a kind of biomass equivalent of dominating blue-green species in lake Müggelsee. On the basis of this series it is possible to estimate relations between an important biological state parameter of the lake and climatological parameters such as global radiation, precipitation or types of zonal atmospheric circulation.

The statistical analysis is realized for seasonal series, using the following cumulative values for each year:

- global radiation in Potsdam from March to June (a longperiodic variation is substracted by global radiation (see OLBERG [8])),
- precipitation at Fürstenwalde/Spree from December to May,
- zonal circulation forms in Middle Europa after HESS/BREZOWSKY from May to October, and
- buoyant part of seston wet volume from May to October.

The character of investigations is diagnostical, i.e., our aim is the estimation of peridic fluctuations, their changes in time and coherence relations. This task can be solved using the following statistical methods: univariate autoregressive spectral estimation, multichannel autoregressive coherence analysis and numerical filtering (see OLBERG and RÁKOCZI [10]; OLBERG and v. SCHOENERMARK [11]; OLBERG [9]).

3. Results

The application of the one channel autoregressive analysis to the buoyant part of seston shows several peaks in the spectrum (see Fig. 2). Peaks appear at periods of about 2.2, 5 and 24 years. Because the sample size of our series (48 values) is small, the existence of the longest period is insecure. Therefore, this period was disregarded for additional analysis. However, the published values of phytoplankton total cell numbers in the Lake Michigan (BROOKS et al. [3]) are likely to show such a long periodicity, too. Accordingly, we are entitled to believe that a period of about 24 years should not be attributed to statistical effects. The spectra of the cited climatological series signal the existence of periods of about 2.2 and 5 years (Fig. 2).

This means that the structure of the one-channel spectra of all analysed series resemble each other.

Similar structures were found in other time series of water quality parameters. For instance, AHL [1] analysed the series of total hardness, bicarbonate, $KMnO_4$-consumption and colour in the Lake Mälaren from 1943 to 1975. He found periods between 6 and 7 years and one year. Only the series of bicarbonate showed different periodicities of about 2 and 4 years. AHL pointed out that especially the spectral density peak at 6 to 7 years corresponds to a peak in the water level spectrum.

ZIMBALEVSKAYA [15] referred to a correspondence between long-term cycles of

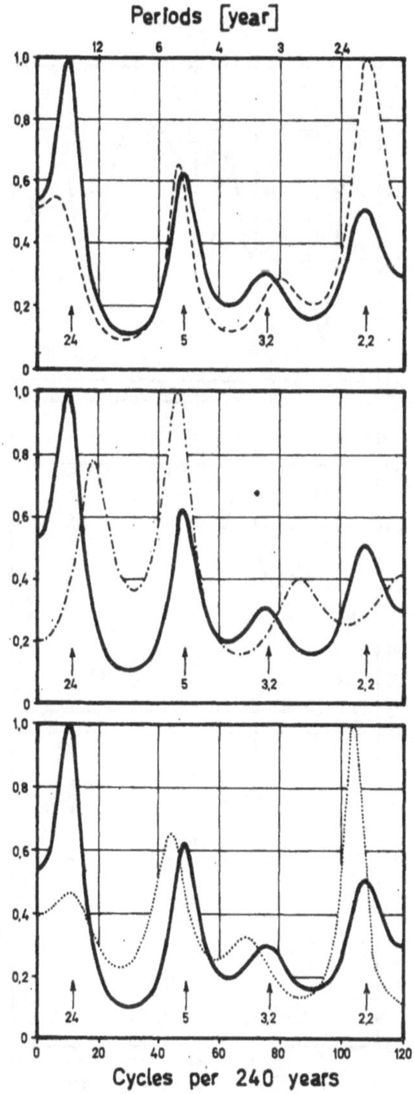

Fig. 2. Maximum-entropy-spectra (MES) for buoyant part of seston content (————), global radiation (-----), precipitation (—·—·—·), and zonal circulation (············). The MES-values are depicted relative to their maximum value

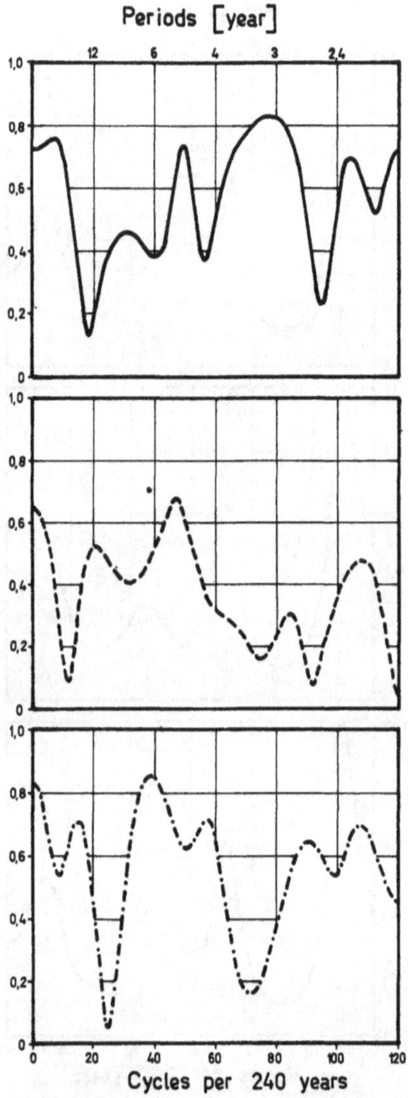

Fig. 3 Coherence specta between the time series of buoyant part of seston content and global radiation (————), buoyant part of seston content and precipitation (- - - -), buoyant part of seston content and zonal circulation (—·— —·)

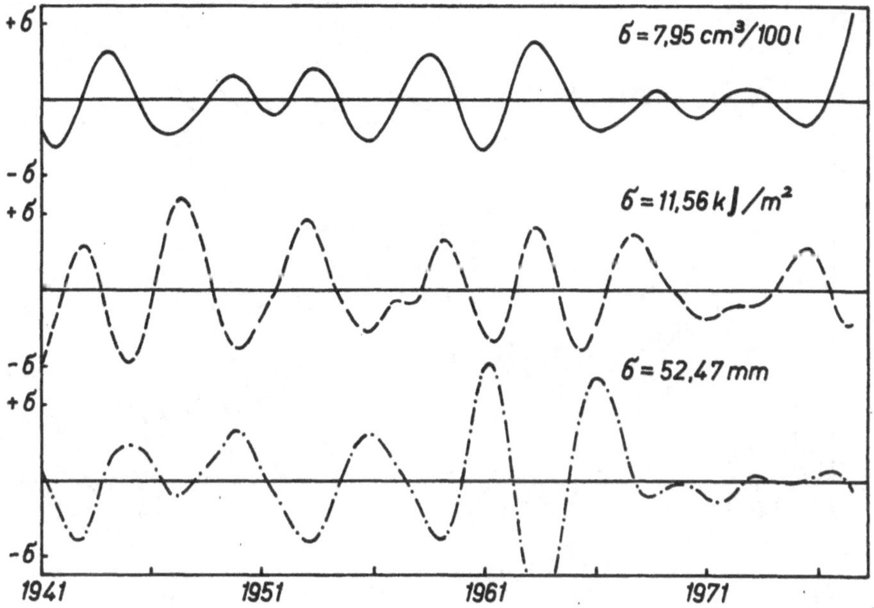

Fig. 4. Results of numerical filtering of the time series in the frequence band of a nearly 5 year period. (buoyant part of seston content: ———, global radiation: ----, and precipitation: —·—·—)

water quality parameters and hydrometeorological or climatic factors, respectively.

The existence of relations between the main periodicities in the buoyant part of seston content and climatological parameters was confirmed by a multi-channel autoregressive spectral analysis. A good correlation appears between buoyant part of seston content and global radiation for the two frequency bands (2,2 and 5 years), but a phase lag is visible (Fig. 3). This phase lag was approximately one year on an average between series of global radiation and seston content such that the maximum of global radiation was past before the maximum of seston content. This phenomen may be shown distinctly by means of numerical filtering (Fig. 4 and 5). At present it is not possible to explain this relationship. For this purpose it is necessary to enlarge the investigations to finding the causes of such phenomenon and to repeat the statistical studies for other time periods, using still other quality parameters.

One reason for this relation could be the ability of such blue-green species to form permanent cells which are able to hibernate in the sediment.

So, the results of large-scale studies differed from those found by short-range analysis. There are no phase lags. Regarding the curves of numerical filtering, it

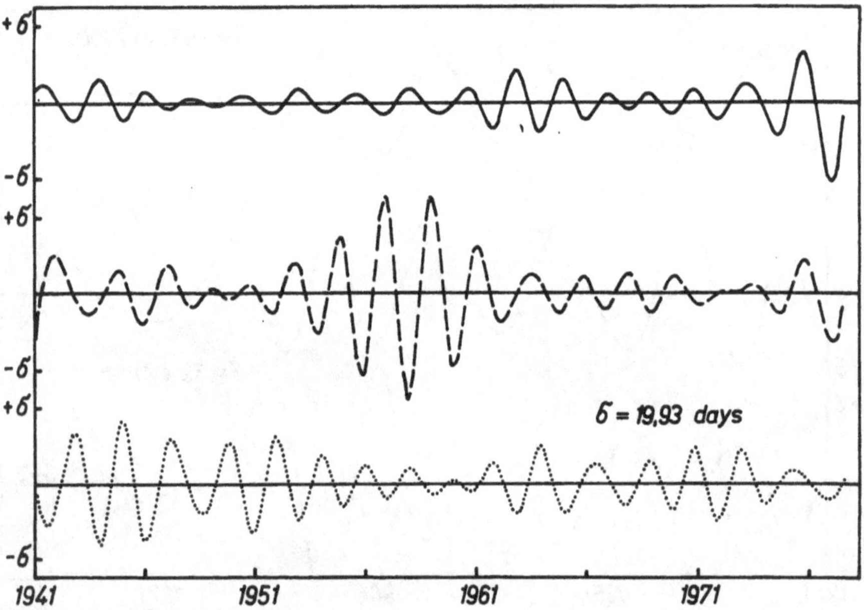

Fig. 5. Results of numerical filtering of the time series in the frequence band of a nearly 2 year period. (buoyant part of seston content: ————, global radiation: – – – – – –, and zonal circulation: ··············).

seems that just after 1970 the amplitudes and also the lag have changed. One reason for this violent change of the ecosystem may be the rapid eutrophication of the lake in that time. This period was characterized by an increase of the sinking part of seston and the complete disappearence of submersed macrophytes in the lake (BARTHELMES [2]). Fig. 4 and 5 show this as well. At the same time, the global radiation was decreasing. On the other hand, the filtered series end with 1971 or 1975, respectively, because the filtering procedure decrease the sample size, and so the recent development cannot be estimated.

Supposing meteorological conditions and eutrophication to be important factors influencing the quality changes of an aquatic ecosystem, the following action chain could be imaginable: Decrease of global radiation causes a decrease in growth of submersed macrophytes — this effects a minor nutrient uptake by submersed macrophytes and so an increase of nutrient content in the open water — this gives rise to forced growth of several phytoplankton species, especially of Diatoms, which increase the water turbidity and decrease the effective light climate in the water body, . . .

The here presented first analyses of long-term changes of a biological ecosystem parameter and its connection with changes of climatological or meteorological parameters show, on the one side, the possibilities but also the limitations existing in this field and on the other hand, emphasize the necessity to enlarge studies on the long-term behaviour of ecosystems and ecosystem parameters.

Acknowledgement

We are indebted to colleagues of the water works Berlin-Friedrichshagen for collecting and delivery of data of seston wet volume of lake Müggelsee.

References

[1] AHL, T.: Natural and human effects on trophic evolution. — Arch. Hydrobiol. Beih. Ergebn. Limnol. **13** (1979), 259—277.

[2] BARTHELMES, D.: Langfristige, biologische Veränderungen im Großen Müggelsee (Berlin), einige Kennzahlen des jetzigen Zustandes und bisherige fischereiliche Auswirkungen. I. Makrophyten. — Z. Binnenfischerei DDR **25** (1978), 171—175.

[3] BROOKS, A. S.; WARREN, G. J.; BORAAS, M. E.; SCALE, D. B.; EDGINTON, D. N.: Long-term phytoplankton changes in Lake Michigan: Cultural eutrophication or biotic shifts. — Verh. Internat. Verein. Limnol. **22** (1984), 452—459.

[4] CHELTON, D. B.; BERNAL, P. A.; MCGOWAN, J. A.: Large-scale interannual physical and biological interaction in the California Current. — J. Mar. Res. **40** (1982) 4, 1095—1125.

[5] HOEG, S.: Betrachtungen zur Jahresdynamik des Phytoplanktons im Müggelsee (1976—1979). — Acta hydrophys. **28** (1983), 5—36.

[6] LUND, J. W. G.: Eutrophication. — Proc. R. Soc. Lond. B. **180** (1972), 371—382.

[7] LUND, J. W. G.: Changes in the phytoplankton of an English lake, 1945—1977. — Hydrobiological Journal **14** (1978), 6—21.

[8] OLBERG, M.: Langperiodische Schwankungen der Strahlungsbilanz, des Niederschlages und des Trockenheitsindex in Potsdam. — Z. Meteor. **29** (1979), 50—55.

[9] OLBERG, M.: Statistische Analyse meteorologischer-klimatologischer Zeitreihen. — Abh. MD der DDR **17** (1982) 128, 129—141.

[10] OLBERG, M.; RÁKÓCZI, F.: Informationstheorie in Meteorologie und Geophysik. — Berlin: Akademie-Verlag, 1984.

[11] OLBERG, M.; v. SCHOENERMARK, M.: Numerische Experimente mit Kohärenz- und gleitenden Maximum-Entropie-Spektren. — Z. Meteor. **33** (1983), 111—115.

[12] RAMBERG, L.: Relations between phytoplankton and light climate in two Swedish forest lakes. — Int. Rev. ges. Hydrobiol. **64** (1979), 749—782.

[13] ROBINSON, G. A.: Continuous plankton records: Phytoplankton in the North Sea, 1958—1980, with special reference to 1980. — Br. Phycol. J. **18** (1983), 131—139.

[14] RUSSEL, F. S.:A summary of the observations on the occurence of planktonic stages of fish off Plymouth 1924—1972. — J. Mar. biol. Ass. U.K. **53** (1973), 347—355.

[15] ZIMBALEVSKAYA, L. N.: Successions, monitoring and predictions of water ecosystems. — Hydrobiol. J. **21** (1985), 3—9.

Acta hydrophys., Berlin **31** (1987) 3/4, S. 153—163

Coastal Research in the GDR
Bibliography 1983—1985

by R. LAMPE[1])

Summary: This bibliography continues the series of literature compilations in this periodical concerning the coast and coastal waters of the GDR. It contains all scientific papers published in the GDR between 1983 and 1985.

Zusammenfassung: Die vorliegende Bibliographie setzt die Reihe der Literaturzusammenstellungen zu Problemen der Küste und der Küstengewässer der DDR, die bisher in dieser Reihe erschienen sind, fort. Sie umfaßt alle wissenschaftlichen Arbeiten, die im Zeitraum 1983—1985 in der DDR publiziert wurden.

In this periodical some bibliographies about the coastal literature are published in the last years (see Acta hydrophys. **6** (1959/60) 2, 73—96; **8** (1963/64) 2, 83—96; **13** (1968/69) 3, 121—136; **19** (1974/75) 1, 29—48; **25** (1980) 4, 333—349; **30** (1986) 4, 213—229). Now this series will be continued. All scientific papers are represented that affects the coast or the coastal waters of the German Democratic Republic and are published between 1983—1985. State-standards or non-published scientific studies, thesises and similars are disregarded. Also publications appears in foreign journals are only considered if they arise from international co-operation.

For easier discovery of thematic specialized titles the bibliography is divided into sections, although this implies some difficulties by setting the article into the right place. If there are some possibilities of an arrangement the literal sense of title decided. So each paper is cited only one time.

At last I am indebted to Mrs. FRIEDRICH and Mr. SCHÜMANN for their help.

[1]) Dr. rer. nat. REINHARD LAMPE, Ernst-Moritz-Arndt-Universität, Sektion Geographie, Friedrich-Ludwig-Jahn-Straße 16, Greifswald, DDR-2200.

1. Common papers

[1] HAHLBECK, W.-H.: Technologie und Technik der intensiven Fischproduktion in seegebundenen Anlagen in Küstengewässern. — Fischerei-Forsch. **21** (1983) 1, 35—45.

[2] HAHLBECK, W.-H.: Eine Futterboot-Technologie für schwimmende Netzkäfiganlagen in Küstengewässern der DDR. — Fischerei-Forsch. **23** (1985) 2, 15—20.

[3] HAHLBECK, W.-H.: Geräteentwicklungen für die Speiseforellenaufzucht in den Küstengewässern der DDR. — Z. Binnenfisch. DDR **32** (1985) 8, 239—244.

[4] HAHLBECK, W.-H.: Einige gegenwärtige Aspekte der Technologien und Technik der Regenbogenforellenmast in den Küstengewässern der DDR. — Seewirtschaft **17** (1985) 9, 453.

[5] HERING, G.: Die Nutzung von erwärmtem Kühlwasser im Küstenbezirk der DDR. — Z. Binnenfisch. DDR **31** (1984) 4, 101—105.

[6] KLIEWE, H.: Untersuchungen über naturräumliche Strukturen und Prozesse im Jungmoränen- und Küstengebiet der DDR als Grundlage für landeskulturelle Maßnahmen. Eine fünfjährige Forschungsbilanz. — Wiss. Z. Univ. Greifswald, math.- nat. R. **32** (1983) 1/2, 3—20.

[7] KLIEWE, H.; LANGE, D.: Fortschritte der Ostsee- und Ostseeküstenforschung.— Arbeitstagung „Weltozean" in Rostock-Warnemünde 1984. — Peterm. Geogr. Mitt. **129** (1985) 4, 263—264.

[8] MATTHÄUS, W.: Grundzüge der regionalen und physikalischen Ozeanologie der Ostsee. — Geogr. Ber. **30** (1985) 1, 1—15.

[9] NEHRING, D.: Charakteristik der chemischen und biologischen Ozeanologie der Ostsee. — Geogr. Ber. 30 (1985) 1, 17—28.

[10] POSSELT, U.: Die Entwicklung und der Einsatz von schwimmenden ringförmigen Netzkäfiganlagen für die industriemäßige Forellenproduktion in geeigneten Küstengewässern der DDR. — Fischerei-Forsch. **21** (1983) 1, 46—49.

[11] POSSELT, U.: Entwicklung der Netzkäfiganlage Typ „Rostock" für die Forellenproduktion in den inneren Seegewässern der DDR. — Z. Binnenfisch. DDR **30** (1983) 10, 291—295.

[12] POSSELT, U.; MOHR, TH.: Zusatzausrüstung zum Käfigträger Typ „Rostock" und dessen Nutzung für die Überwinterung von Satzfischen in geschützten Küstengewässern. — Fischerei-Forsch. **23** (1985) 4, 96—100.

[13] SCHIER, N.: Stand und Aufgaben der Hochsee- und Küstenfischerei der DDR. — Z. Binnenfisch. DDR **30** (1983) 3, 99—103.

[14] SCHMAGER, B.: Anforderungen, Hauptaufgaben sowie einige Probleme der angewandten modernen maritimen Meteorologie. — Abh. Meteorol. Dienstes DDR **18** (1985) 134, 89—97.

[15] SCHOSS, M.: 3. Symposium zur Bewirtschaftung der Küstengewässer. — Z. Binnenfisch. DDR **31** (1984) 11, 336—342.

[16] SCHOSS, M.: 4. Symposium zur intensiven Fischproduktion und Bewirtschaftung der Küstengewässer. — Z. Binnenfisch. DDR **32** (1985) 9, 274—281.

[17] STRIGGOW, K.: 25 Jahre Institut für Meereskunde — Die Entwicklung seines wissenschaftlichen Profils. — Beitr. Meereskd. 48 (1983), 5—7.

[18] STRIGGOW, K.: Brachte die „Practical Salinity Scale 1978" eine neue Maßeinheit, eine neue Definition oder eine neue Meßmethode für den Salzgehalt des Meerwassers. — Seewirtschaft **17** (1985) 5, 253—258.

[19] Succow, M.: Moorbildungstypen des südbaltischen Raumes. — Peterm. Geogr. Mitt., Erg.-Heft Nr. 282 (1983), 86—107.

2. Physics of hydro- and atmospheres. Hydrography. Meteorology

[20] Apel, E.: Windwellen auf stehenden Gewässern. — Bemerkungen zu Kommentar und Berechnungsgrundlagen des Standards TGL 28724 und des Werkstandards WAPRO 3.04. — Wasserwirtschaft Wassertechnik **35** (1985) 8,189—190.

[21] Dahms, P.: Fremdwasser in Fluß- und Küstenniederungen. — Wiss. Z. Univ. Rostock, math.-nat. R. **33** (1984) 5, 72—74.

[22] Foken, Th.: Zur Parametrisierung der turbulenten Energieaustauschströme über dem Meer. — Abh. Meteorol. Dienstes DDR **18** (1985) 134, 85—89.

[23] Francke, E.: Ergebnisse langzeitiger Strömungsmessungen der Deckschicht des Seegebietes der Darßer Schwelle. — Beitr. Meereskd. 48 (1983), 23—45.

[24] Georgi, F.: Verteilung und Beschaffenheit des Sestons in inneren Küstengewässern der DDR (Darß-Zingster Boddengewässer) sowie der westlichen und mittleren Ostsee unter besonderer Berücksichtigung der Wasseraustauschprozesse zwischen beiden Systemen. — Beitr. Meereskd. 52 (1985), 35—48.

[25] Hupfer, P.: Wechselwirkung zwischen Meer und Atmosphäre unter den Bedingungen unmittelbarer Küstennähe. — Geodät. — geophysik. Veröff., R. IV, H. 38 (1983), 3—21.

[26] Hupfer, P.: Zur Abschätzung der Windströmungen in der ufernahen Zone der Ostsee.— Geodät.- geophysik. Veröff., R. IV, H. 38 (1983), 55—66.

[27] Kremser, U.: Eine Bemerkung zur Rolle der Turbulenz in hydrologischen Prozessen. — Wiss. Z. Univ. Rostock, math.-nat. R. **33** (1984) 5, 78—79.

[28] Kuznecov, C.A.; Raabe, A.: Experimentelle Untersuchung der horizontalen Heterogenität des Windfeldes in der Küstenzone und Probleme ihrer theoretischen Beschreibung. — Acta hydrophys. **29** (1985) 2/3, 153—163.

[29] Matthäus, W.: Zur mittleren jahreszeitlichen Veränderlichkeit von Temperatur und Salzgehalt in der Mecklenburger Bucht. — Beitr. Meereskd. 50 (1984), 9—23.

[30] Matthäus, W.; Laas, H. U.; Francke, E.; Schwabe, R.: Zur Veränderlichkeit des Volumen- und Salztransportes über die Darßer Schwelle. — Gerlands Beitr. Geophys. **92** (1983) 5, 407—420.

[31] Panin, G. N.; Karpovič, Ju. V.; Kusnecov, O. A.; et al.: Resul'taty spravnenija priborov ismerjajuščich fluktuacii skorosti, temperatury i vlažnosti vozducha. — Z. Meteorol. **35** (1985) 2, 84—89.

[32] Prandke, H.: Mikrostrukturuntersuchungen in der Ostsee. — Seewirtschaft **15** (1983) 2, 91—96.

[33] Raabe, A.: Zur Wechselwirkung von Meer und Atmosphäre in Küstennähe unter Berücksichtigung der internen Grenzschicht im Windfeld der atmosphärischen Bodenschicht. — Geodät. — geophys. Veröff., R. IV H. 38 (1983), 22—54.

[34] Raabe, A.: On the relation between the drag coefficient and fetch above the sea in the case of off-shore wind in the near-shore zone. — Z. Meteorol. **33** (1983) 6, 363—367.

[35] Tiesel, R.: Die Wärmezyklonen der westlichen und mittleren Ostsee. — Z. Meteorol. **37** (1984) 6, 354—365.

[36] Tremp, D.; Thieme, A.: Ermittlung von Wasserstandsganglinien im seeartig erweiterten Gebiet des Peenestromes bei Ostseehochwasser. — Wiss. Z. Univ. Rostock, math.-nat. R. **33** (1984) 5, 66—68.

[37] Wagner, H.: Empirisch-rechnerische Prognose von durch Wind erzeugten Wellen. — Seewirtschaft **17** (1985) 6, 298—302.

3. Chemistry

[38] Georgi, F.; Nausch, G.: Untersuchungen zur Veränderlichkeit chemischer Parameter des Sestons aus flachen Küstengewässern. — Wiss. Z. Univ. Rostock, math.-nat. R. **33** (1984) 6, 56—58.

[39] Nausch, G.: Phosphoreintrag aus Niederschlägen im Küstengebiet der DDR. — Acta hydrochim. hydrobiol. **11** (1983) 1, 151—154.

[40] Nausch, G.; Schlungbaum, G.: Sedimentchemische Untersuchungen in Küstengewässern der Deutschen Demokratischen Republik. — Teil 17: Spezielle Untersuchungen zur Dynamik der Oberflächensedimente eines flachen Boddengewässers (Barther Bodden). — Acta hydrochim. hydrobiol. **12** (1984) 1, 61—72.

[41] Nausch, G.; Schlungbaum, G.: Sedimentchemische Untersuchungen in Küstengewässern der DDR. — Teil 20: Die Sedimente der Darß-Zingster Boddengewässer — eine zusammenfassende und vergleichende Übersicht. — Wiss. Z. Univ. Rostock, math.-nat. R. **33** (1984) 6, 59—63.

[42] Nehring, D.; Francke, E.: Die hydrographisch-chemischen Bedingungen in der Ostsee im Jahre 1981. — Fischerei-Forsch. **21** (1983) 1, 45—55.

[43] Nehring, D.; Francke, E.: Die hydrographisch-chemischen Bedingungen in der Ostsee im Jahre 1982. — Fischerei-Forsch. **21** (1983) 1, 56—65.

[44] Nehring, D.; Francke, E.: Die hydrographisch-chemischen Bedingungen in der westlichen und zentralen Ostsee im Jahre 1983. — Fischerei-Forsch. **23** (1985) 4, 7—17.

[45] Nehring, D.; Francke, E.: Die hydrographisch-chemischen Bedingungen in der westlichen und zentralen Ostsee im Jahre 1984. — Fischerei-Forsch. **23** (1985) 4, 18—27.

4. Geology. Geography

[46] Brügmann, L.; Lange, D.: Geochemische und sedimentologische Untersuchungen an einem Sedimentkern aus dem Schlickgebiet der Lübecker Bucht. — Gerlands Beitr. Geophys. **92** (1983) 2/4, 241—268.

[47] Gellert, J. F.: Begriff und Wesen des Strandes als geomorphologische und naturräumliche Erscheinung. — Wiss. Z. Päd. Hochsch. Potsdam, naturwiss. R. **27** (1983) 3, 443—449.

[48] Gellert, J. F.: Organische Strandwälle und Strandplatten (Thanotope) auf dem Seestrand der Schaabe (Nordrügen) als aktuogeologische Erscheinung. — Z. geol. Wiss. **13** (1985) 4, 533—536.

[49] Gellert, J. F.: Strukturen und Prozesse am Meeresstrand als geomorphologischem und landschaftlichem Grenzsaum zwischen Land und Meer — geologisch-geographische Analyse und Synthese. — Peterm. Geogr. Mitt. **129** (1985) 4, 239—252.

[50] GELLERT, J. F.; HEYER, E.; NEUMEYER, G.: Ergebnisse morphographischer und morphodynamischer Untersuchungen am Seestrand der Schaabe (Nord-Rügen). — Acta hydrophys. **29** (1985) 2/3, 93—152.

[51] GURWELL, B. R.: Beitrag zur uferparallelen Variation sedimentpetrographischer Merkmale. — Z. geol. Wiss. **11** (1983) 1, 59—76.

[52] GURWELL, B. R.: Komplexe Bearbeitung der Steiluferproblematik und erste Ergebnisse vom Hohen Ufer des Fischlandes. — Wiss. Z. Univ. Greifswald, math.- nat. R. **34** (1985) 4, 33—36.

[53] GURWELL, B. R.; ZIELISCH, E.: Beitrag zur Messung der Sedimentdynamik im Riffbereich. — Z. geol. Wiss. **11** (1983) 2, 209—223.

[54] GUSEN, R.: Der lithologische Bau der Schorre von Neu Reddevitz (Greifswalder Bodden).— Z. geol. Wiss. **11** (1983) 2, 193—208.

[55] JANKE, W.: Verbreitung und Eigenschaften typischer Sedimente küstennaher Talungen und Niederungen unter besonderer Berücksichtigung anthropogen bedingter Einflüsse. — Peterm. Geogr. Mitt., Erg.-Heft Nr. 282 (1983), 118—126.

[56] KLIEWE, H.: Entwicklung, Ergebnisse und Tendenzen der Spätglazial- und Holozänforschung im nordöstlichen Küstenraum der DDR. — Peterm. Geogr. Mitt., Erg.-Heft Nr. 282 (1983), 28—38.

[57] KNOLL, B.: Sedimentverteilung und Sedimentumlagerung auf der Schorre vor Neu-Reddevitz (Südost-Rügen). — Wiss. Z. Univ. Greifswald, math.- nat. R. **32** (1983) 1/2, 99—108.

[58] KNOLL, B.: Der Uferrückgang des Neu-Reddevitzer Kliffs. — Wiss. Z. Univ. Greifswald, math.-nat. R. **32** (1983) 1/2, 109—112.

[59] KOLP, O.: Die schrittweise Verlagerung der Odermündung von der Bornholmmulde bis in die Oderbucht infolge holozäner Wasserstandsänderungen im südlichen Ostseeraum. — Peterm. Geogr. Mitt. **127** (1983) 2, 73—87.

[60] KOLYSCHKOV, P.; PESCHEL, G.: Untersuchung der Zusammenhänge zwischen Eigenschaften geologischer Objekte mit Hilfe nichtlinearer Regressionsbeziehungen. — Z. angew. Geol. **31** (1985) 7, 175—178.

[61] LAMPE, R.: Die Sedimente eines polytrophen Flachsees (Schmachter See/Rügen) und ihre Beziehungen zum Stoffhaushalt in Vergangenheit und Gegenwart. — Wiss. Z. Univ. Greifswald, math.-nat. R. **32** (1983) 1/2, 93—98.

[62] LANGE, D.; NIEDERMEYER, R.-O.: Zur Untersuchung von Sedimentationsprozessen in der westlichen Ostsee (Mecklenburger Bucht) — Methodik und erste Ergebnisse. — Wiss. Z. Univ. Greifswald, math.-nat. R. **34** (1985) 4, 27—30.

[63] OTTO, G.: Strukturveränderungen in holozänen Küstenniederungen durch Komplexmelioration, dargestellt am Beispiel der Conventer Niederung bei Bad Doberan. — Peterm. Geogr. Mitt., Erg.-Heft Nr. 282 (1983), 72—84.

[64] PESCHEL, G.; WALTHER, K.-F.: Lithofazielle Charakterisierung des Meeresbodensediments durch Analyse von Sondenschleppgeräuschen. — Z. angew. Geol. **29** (1983) 1, 38—42.

[65] ROGGE, H.-J.; WALTHER, K.-F.; GROMOLL, L.: Beiträge des Zentralen Geologischen Institutes zur Suche und Erkundung von Kiessand-Lagerstätten auf dem Festlandsockel der DDR. — Z. angew. Geol. **31** (1985) 12, 300—304.

[66] ROTHER, K.: Paläomagnetische Untersuchungen im Quartär der Ostsee — erste Ergebnisse. Z. angew. Geol. **31** (1985) 11, 289—290.

[67] SCHUMACHER, W.: Strukturelle Sedimentmerkmale als Seespiegelindikatoren. — Wiss. Z. Univ. Greifswald, math.-nat. R. **34** (1985) 4, 30—32.

[68] STRIGGOW, K.: Die relative Landsenkung im Bereich des Sundes und der Beltsee — eine weitere Ursache der rezenten Salzgehaltszunahme der Ostsee. — Gerlands Beitr. Geophys. **92** (1983) 2/4, 228—240.

[69] WELLER, H.: Charakterisierung von Korngrößenverteilungen durch Momentparameter. — Wiss. Z. Univ. Greifswald, math.-nat. R. **34** (1985) 4, 13—16.

5. Biology

5.1. Hydrobiology

[70] ARNDT, E. A.: The ecological niche of *Cordylophora caspia* (PALLAS 1771). — Limnologica **15** (1984) 2, 469—477.

[71] ARNDT, H.; HEERKLOSS, R.; SCHNESE, W.: Seasonal and spatial fluctuations of estuarine rotifers in a Baltic inlet. — Limnologica **15** (1984) 2, 377—385.

[72] HEERKLOSS, R.; ARNDT, H.; HELLWIG, J.; et al.: Consumption and assimilation by zooplankton related to primary production in the Baltic coastal water inlet Barther Bodden. — Limnologica **15** (1984) 2, 387—394.

[73] HÜBENER, TH.; GASE, G.; GENČIOVA, A.; et al.: Biomasse, Primärproduktion und Sukzessionsfolge des Phytoplanktons der Unterwarnow. — Wiss. Z. Univ. Rostock, math.-nat. R. **33** (1984) 6, 43—52.

[74] JANSEN, W.: Fang und Verhalten von *Neomysis integer* LEACH (Crustacea, Mysidacea) in der Darß-Zingster Boddenkette (südliche Ostsee). — Fischerei-Forsch. **23** (1985) 4, 33—36.

[75] JANSEN, W.: Biometrische Grunddaten von *Neomysis integer* LEACH aus der Darß-Zingster Boddenkette (südliche Ostsee). — Fischerei-Forsch. **23** (1985) 4, 37—40.

[76] JANSEN, W.: Stellung von *Neomysis integer* LEACH (Crustacea, Mysidacea) als Konsument im Nahrungsgefüge der Darß-Zingster Boddenkette (südliche Ostsee). — Fischerei-Forsch. **23** (1985) 2, 55—59.

[77] JONCZYK, E.; RADZIEJEWSKA, T.: Temporal changes in sand meiofauna of a southern Baltic beach. — Limnologica **15** (1984) 2, 421—423.

[78] JOST, G.; GÜNTHER, B.: Seasonal changes in bacteriological parameters in the chain of shallow water (boddens) south of the Darß-Zingst peninsula (South Baltic). — Limnologica **15** (1984) 2, 597—603.

[79] KELL, V.: Primärproduktionsmessungen im Greifswalder Bodden in den Jahren 1981 und 1982. — Wiss. Z. Univ. Rostock, math.-nat. R. **33** (1984) 6, 53—55.

[80] KÖHLER, S.; ARLT, G.: The distribution of ostracods in an inlet of the Darß-Zingst Bodden chain. — Limnologica **15** (1984) 2, 415—419.

[81] LIBBERT, E.; WALTHER, TH.: Photosynthetic production of a brackish water community of *Chara tomentosa* L. and its dependence on environmental conditions. — Int. Rev. ges. Hydrobiol. **70** (1985) 3, 359—368.

[82] MARANOVA, L.: Notes on water-borne micromycetes in northern parts of the German Democratic Republic. — Feddes Repertorium **95** (1984) 3, 201—207.

[83] RADZIEJEWSKA, T.: Meiofauna communities in organically enriched estuarine environment. — Limnologica 15 (1984) 2, 425—427.

[84] SCHARF, E. M.; SCHNESE, W.: Abundance and reproduction potential of benthic ciliates in a β-mesosaprobic aquatorium. — Limnologica 15 (1984) 2, 429—437.

[85] SCHIEWER, U.: The ecophysiology of salt adaption in fresh-water blue-green algae. — Limnologica 15 (1984) 2, 555—558.

[86] SCHIEWER, U.: In vitro primary productivity (assimilation number) of phytoplankton from the Darß-Zingst estuary. — Limnologica 15 (1984) 2, 575—579.

[87] SCHULZ, S.; BREUEL, G.: On the variability of some biological parameters in the summer pelagic system of the Arkona Sea. — Limnologica 15 (1984) 2, 365—370.

[88] SUBKLEW, H.-J.: Ökologische Beobachtungen bei der biologischen Prüfung von Schiffsbodenfarben. — Acta hydrochim. hydrobiol. 11 (1983) 3, 361—369.

[89] SUBKLEW, H.-J.: Zur Hydrobiologie des Strelasund. — Limnologica 16 (1984) 1, 191—205.

[90] VIETINGHOFF, U.; ERDMANN, N.; ARNDT, H.; et al.: Integrated samples provide accurate means of parameters characterising aquatic ecosystems. — Int. Rev. ges. Hydrobiol. 69 (1984) 1, 121—131.

[91] WACHENFELDT, T. V.: Growth rate of some Baltic macroalgae. — Limnologica 15 (1984) 2, 311—317.

5.2. Biology of fishes

[92] BAST, H.; WINKLER, H.; HAHN, W.: Bemerkungen zur Biologie und Bedeutung des Kaulbarsches (*Gymnocephalus cernua*) der Darß-Zingster Boddenkette. — Fischerei-Forsch. 21 (1983) 4, 34—38.

[93] BRIELMANN: Abundanz und Wachstum der Larven des Rügenschen Frühjahrsherings (*Clupea harengus* L.) aus dem Greifswalder Bodden 1977—1979. — Fischerei-Forsch. 21 (1983) 3, 32—38.

[94] DEBUS, L.; ARNDT, H.: Nahrungsbiologische Untersuchungen an Jungfischpopulationen eines brackigen Flachwassergebietes des Barther Boddens (südliche Ostsee). — Wiss. Z. Univ. Rostock, math.-nat. R. 33 (1984) 6, 76—82.

[95] GIERER, W.: Differentialdiagnostische Untersuchungen von bakteriell kontaminierten Organproben erkrankter Fische aus den Forellenanlagen der Ostseeküste. — Fischerei-Forsch. 21 (1983) 1, 50—51.

[96] GIERER, W.: Ein diagnostisches Programm zur Differenzierung fischpathogener Bakterien aus fischereilichen Produktionsanlagen der Ostseeküste. — Fischerei-Forsch. 23 (1985) 1, 95—96.

[97] JÖNSSON, N.; SCABELL, J.; GENTZEN, B.: Praxisorientierte Verhaltensuntersuchungen am Rügenschen Frühjahrshering (*Clupea harengus* L.). — Wiss. Z. Univ. Rostock, math.-nat. R. 33 (1984) 6, 68—75.

[98] KLINKHARDT, M.: Zum Einfluß des Salzgehaltes auf die Befruchtungsfähigkeit des Laiches der Rügenschen Frühjahrsheringe. — Fischerei-Forsch. 22 (1984) 1, 73—75.

[99] KLINKHARDT, M.; BIESTER, E.: Erste Ergebnisse von experimentellen Felduntersuchungen am Laich von Rügenschen Frühjahrsheringen — Fischerei-Forsch. 22 (1984) 1, 76—78.

[100] KLINKHARDT, M.; LILL, D. R.; BIESTER, E.: Untersuchungen zum Ablauf der Frühjahrs-

laichsaison 1982 und 1983 der Rügenheringe (*Clupea harengus* L.) auf einem ausgewählten Laichplatz des Greifswalder Boddens. — Fischerei-Forsch. **23** (1985) 4, 41—48.

[101] KLINKHARDT, M.; NICOLAUS, G.: Untersuchungen zur Veränderung der Trockenmasse und des Energiegehaltes sich entwickelnder Eier des Rügenschen Frühjahrsherings. — Wiss. Z. Univ. Rostock, math.-nat. R **33** (1984) 6, 64—67.

[102] NOACK, B.: Besatzversuche mit Marmor- und Silberkarpfen (*Aristichthys nobilis* RICH., *Hypophthalmichthys moltrix* VAL.) im Kleinen Jasmunder Bodden, einem stark eutrophen, brackigen Gewässer der südlichen Ostsee. — Fischerei-Forsch. **23** (1985) 2, 44—50.

[103] REIMER, L. W.: Die Rolle der Parasiten bei der Fischaufzucht im Brackwasser der Ostsee. — Fischerei-Forsch. **23** (1985) 1, 97—98.

[104] SPANNHOF, L.: Zu einigen Problemen der Physiologie von Regenbogenforellen unter Brackwasserbedingungen. — Fischerei-Forsch. **23** (1985) 1, 96—97.

[105] SPITTLER, P.; KELL, V.; MÜLLER, S.: Zur Nahrungsselektion von Silberkarpfen und Hybriden in einem eutrophen Brackgewässer. — Fischerei-Forsch. **21** (1983) 1, 60—61.

[106] SPITTLER, P.; WOLTHUSEN, E.; KELL, V.; MÜLLER, S : Nahrungsbiologische Untersuchungen an Silberkarpfen und Marmorkarpfenhybriden unter besonderer Berücksichtigung der Nutzung von sedimentiertem Material. — Wiss. Z. Univ. Rostock, math.-nat. R. **33** (1984) 6, 88—92.

[107] WINKLER, H.; DIBBERN, ST.; JARMATZ, K.; DEBUS, L.: Einige Angaben zum Jungfischaufkommen in ausgewählten Gebieten der Darß-Zingster Boddenkette im Zeitraum von 1981—1983. — Wiss. Z. Univ. Rostock, math.-nat. R. **33** (1984) 6, 83—87.

5.3. Mathematical modelling of aquatic ecosystems

[108] VIETINGHOFF, U.: Felduntersuchungen, Laborexperimente und mathematische Modellierung am Beispiel der Ökosystemanalyse Boddenkette West. — Kolloqu. Theorie — Modell — Experiment — Praxis. — Rostock: Wilh.-Pieck-Univ., 1982, 107—120.

[109] VIETINGHOFF, U.: Mathematic modelling of the ecosystem Barther Bodden. — Limnologica **15** (1984) 2, 253—262.

[110] VIETINGHOFF, U.; HUBERT, M.-L.; HEERKLOSS, R.; et al.: A mathematical submodel for the planktonic Rotatoria in the ecosystem of the Barther Bodden (Southern Baltic Sea). — Int. Rev. ges. Hydrobiol. **69** (1984) 2, 159—172.

5.4. Biology of coastal zone

[111] JESCHKE, L.: Landeskulturelle Probleme des Salzgraslandes an der Küste. — Naturschutzarb. Mecklenb. **26** (1983) 1, 5—12.

[112] JESCHKE, L.: Vegetationsveränderungen in den Küstenlandschaften durch Massentourismus und Nutzungsintensivierung. — Arch. Naturschutz u. Landschaftsforsch. **25** (1985) 4, 223—236.

[113] KRISCH, H.: Biomasseproduktion und edaphische Verhältnisse von *Bolboschoenus maritimus*-Beständen des Greifswalder Boddens. — Limnologica **16** (1985) 2, 277—285.

[114] SCHMIDT, D.: Die Lebens- unù Wuchsformen der Hydro- und Halophyten im Pleistozängebiet der DDR. — Feddes Repertorium **96** (1984) 4, 307—342.

[115] SELLIN, D.: Das NSG „Peenemünder Haken, Struck und Ruden". — Bericht 1981/82 für das Teilgebiet Struck. — Naturschutzarb. Mecklenb. **26** (1983) 2, 101—102.

[116] SELLIN, D.: Zum Vorkommen des Seggenrohrsängers im Gebiet des Peenestromes und im NSG „Peenemünder Haken, Struck und Ruden". — Naturschutzarb. Mecklenb. **27** (1984) 1, 21—24.

[117] SELLIN, D.: Zum Stieleichenbestand im NSG Struck. — Naturschutzarb. Mecklenb. **28** (1985) 1, 47—49.

[118] SELLIN, D.: Das NSG Peenemünder Haken, Struck und Ruden. — Bericht 1983/84 für das Teilgebiet Struck. — Naturschutzarb. Mecklenb. **28** (1985) 1, 52—53.

[119] SLOBODDA, S.: Zur Struktur und Dynamik von Bodden-Verlandungssäumen an der Ostseeküste der DDR unter Berücksichtigung der aktuellen Nutzungsverhältnisse und aktueller Umweltbelastung. — VII. Internat. Symposium über die Problematik der ökologischen Landschaftsforsch., Bratislava/Pezinok, 21.—26. 10. 1985, Symposiumsband 3.

[120] VOIGTLAND, R.: Wasserchemismus und Biomasseproduktion in Schilfröhrichten und deren Beeinflussung durch Grund- und Boddenwässer. — Wiss. Z. Univ. Greifswald, math.-nat. R. **32** (1983) 1/2, 113—120.

5.5. Environmental protection

[121] CONRAD, U.; TEICHMANN, A.: Untersuchungen zum Schwimmentendurchzug an der Südküste des Greifswalder Boddens. — Ornith. Rundbrief Mecklenb. **26** (1983), 33—34.

[122] DAUBER, M.: Die Brutvögel der Insel Riems (Greifswalder Bodden). — Eine Darstellung der Bestandsentwicklung von 1860—1977. — Beitr. Vogelkd. **29** (1983), 57—76.

[123] DWENGER, R.: Chileflamingos im Ostseebodden. — Falke **30** (1983) 3, 93—94.

[124] FUKAREK, F.: Die Verbreitung der Lycopodiales im Norden der DDR. — Gleditschia **13** (1985) 1, 127—140.

[125] FUKAREK, F.: Rote Liste der verschwundenen und gefährdeten Höheren Pflanzen von Mecklenburg. — 3. Fassung. — Bot. Rundbrief für den Bez. Neubrandenburg **16** (1985), 3—43.

[126] GROTH, K.: Aufgaben und Organisation des Küstenvogelschutzes im Bezirk Rostock. — Naturschutzarb. Mecklenb. **26** (1983) 2, 105—106.

[127] HAMANN, J.: Der Brutbestand der Limikolen in der Wismar-Bucht. — Ornith. Rundbrief Mecklenb. **28** (1985), 9—29.

[128] LEIPE, TH.: Spülfeldprobleme um ein Küstenvogelbrutgebiet im westlichen Greifswalder Bodden. — Naturschutzarb. Mecklenb. **26** (1983) 1, 19—24.

[129] LEIPE, TH.: Zur Nahrungsökologie der Eisente (*Clangula hyemalis*) im Greifswalder Bodden. — Beitr. Vogelkd. **31** (1985) 1/3, 121—140.

[130] LEIPE, TH.; SELLIN, D.: Zum Vorkommen von Bergenten und Eisenten auf dem Greifswalder Bodden. — Ornith. Rundbrief Mecklenb. **26** (1983), 34—47.

[131] PANKOW, H,: Verschollene, gefährdete und interessante Großalgen im nördlichen Gebiet der DDR. — Bot. Rundbrief für den Bez. Neubrandenburg **16** (1985), 65—72.

[132] PLATH, L.: Zur Bestandsgröße der an den Kreideküsten der Insel Rügen brütenden Mehlschwalben (*Delichon urbica*) nebst Bemerkungen zum dortigen Vorkommen der Uferschwalbe (*Riparia riparia*). — Arch. Freunde Naturgesch. Mecklenb. **23** (1983), 5—8.

[133] SCHIEMANN, H.: Die Wassertreter in den Nordbezirken der DDR. — Falke **31** (1984) 3, 94—98.

[134] TEICHMANN, A.; CONRAD, U.: Zum Limikolenzug am südlichen Greifswalder Bodden auf der Grundlage von Planbeobachtungen. — Ornith. Rundbrief Mecklenb. **27** (1984), 8—35.

6. Archaeology. History

[135] GRAMSCH, B.: Archäologische Untersuchungen aus steinzeitlichen Küstensiedlungsplätzen Rügens und ihre Aussagen zur Ostseegeschichte (Zusammenfassung). — Peterm. Geogr. Mitt., Erg.-Heft Nr. 282 (1983), 39—40.

[136] HERRMANN, J.: Ralswiek — Seehandelsplatz, Hafen und Kultstätte. — Arbeitsstand 1983. — Ausgrabungen, Funde **29** (1984) 3, 128—135.

[137] LAMPE, W.: Ueckeritz- — Ein jungbronzezeitlicher Hortfund von der Insel Usedom. — Beitr. Ur- u. Frühgesch. Bez. Rostock, Schwerin, Neubrandenburg, **15** (1982), 103 S.

[138] SCHOMBURG, J.; NILIUS, I.; ZWAHR, H.: Zur Stoffbestandsuntersuchung neolithischer Keramik der Siedlungsplätze Gristow, Kr. Greifswald, und Ralswiek, Kr. Rügen. — Ausgrabungen, Funde **29** (1984) 3, 101—106.

7. Cartography. Geodesy

[139] KRAMER, D.: Einsatz elektronischer Anlagen in der Seevermessung. — Vermessungstechnik **31** (1983) 12, 407—410.

[140] LINKE, K.: Untersuchungen zur Anwendung der Zweimedien-Photogrammetrie für die Bestimmung von Wassertiefen in Flachwassergebieten. — Vermessungstechnik **33** (1985) 10, 346—348.

[141] WEISE, H.: Hydrokinematisches Nivellement. — Vermessungtechnik **33** (1985) 12, 406—409.

8. Coastal protection

[142] ENGELKE, G.; LATTERMANN, E.: Die Standsicherheit von Betondeckwerken auf Böschungen wasserwirtschaftlicher Anlagen. — Wiss. Z. Techn. Univ. Dresden **32** (1983) 1, 150—153.

[143] GURWELL, B. R.; JÄGER, B.: Küstenveränderung und Küstenschutz dargestellt am Beispiel des Abschnittes Dranske/Rügen. — Peterm. Geogr. Mitt. **127** (1983) 1, 15—24.

[144] JÄGER, B.; DITTRICH, H.; WEISS, D.: Untersuchungen von technischen Textilien und ihr Einsatz im Küstenwasserbau der DDR. — Seewirtschaft **16** (1984) 5, 244—249.

[145] MUCHA, A.; SZCZECHOWSKI, B.: Photogrammetrische Vermessung des Wellengangs an hydrotechnischen Modellen. — Jenaer Rdsch. **28** (1983) 4, 198—200.

[146] ROSENLÖCHER, B.: Die Aufgaben des Küstenschutzes. — Wasserwirtschaft, Wassertechnik **35** (1985) 2, 32—34.

[147] WEISS, D.; JÄGER, B.: Das Küstenschutzsystem aus T-Buhnen und einem Natursteinwellenbrecher bei Dranske/Rügen. — Beitr. Meereskd. 49 (1983), 99—137.

9. Instrumentation. Data processing

[148] Czepa, O.: Mikroprozessorgesteuerte Meßwerterfassung. — Acta hydrophys. **28** (1983) 3, 177—186.

[150] Keding, H.-J., Seyboth, P.; Norden, A.: Unterwasserfernseheinrichtung auf der Basis einer TFK 500. — Seewirtschaft **16** (1984) 11, 560—561.

[151] Kiekbusch, A.: Piezoelektrischer Ultraschallwandler für die telemetrische Unterwasserkommunikation. — Seewirtschaft **15** (1983) 12, 607—609.

[152] Peschel, G.: Rechnergestützte Erkundung von Deckgebirgsrohstoffen. — Z. angew. Geol. **31** (1985) 9, 221—228.

[153] Prandke, H.; Krüger, S.; Roeder, W.: Aufbau und Funktion einer frei fallenden Sonde zur Untersuchung der Mikrostruktur der thermohalinen Schichtung im Meer. — Acta hydrophys. **29** (1985) 2/3, 165—210.

[154] Schönfeldt, H.-J.; Raabe, A.; Piazena, H.: Erfahrungen beim Einsatz einer hochempfindlichen Feuchtesonde mit Miniaturfeuchtefühlern auf Bariumfluoridbasis. — Vergleich mit einem Lyman-α-Hygrometer. — Z. Meteorol. **33** (1983) 2, 116—118.

[155] Striggow, K.; Dankert, R.: Induktive Leitfähigkeitssensoren für die ozeanologische Anwendung. — Physikalische Wirkprinzipien, theoretische Grundlagen und die Möglichkeiten zur Eliminierung des Einflusses einer variablen Kernpermeabilität. — Beitr. Meereskd. 50 (1984), 25—41.

[156] Wiemer, R.; Schefter, S.: Rechnereinsatz zur Lösung der Aufgaben des technischen Küstenschutzes. — Wasserwirtschaft, Wassertechnik **35** (1985) 5, 102—104 u. 116.

10. Bibliographies

[157] Bibliographie zur Ur- und Frühgeschichte. — 1. 3. 1982 bis 28. 2. 1983 mit Nachträgen. — Ausgrabungen, Funde **28** (1983) 6, 263—306.

[158] Bibliographie zur Ur- und Frühgeschichte. — 1. 3. 1983 bis 29. 2. 1984 mit Nachträgen. — Ausgrabungen, Funde **29** (1984) 6, 269—307.

[159] Mecklenburgische Bibliographie. — Regionalbibliographie der Bezirke Rostock, Schwerin und Neubrandenburg, Berichtsjahr 1983. — Schwerin: 1985.

[160] Schröder, A.: Gesamtinhaltsverzeichnis zu den Heften 1—50 der „Beiträge zur Meereskunde". — Beitr. Meereskd. 50 (1984), 51—76.

Acta hydrophys., Berlin **31** (1987) 3/4, S. 165—172

Deterministic and Stochastic Phases in the Time Evolution of Water Resources

By P. MAUERSBERGER[1])

Summary: In all time-scales evolution of water resources is based on non-linearity, non-stationarity and threshold processes. Therefore, the theory of water resources systems has to involve deterministic methods in order to deal with processes far from bifurcations as well as stochastic methods and synergetics which offer tools for dealing with the dynamics in the vicinity of bifurcations of the non-linear deterministic differential equations. Not only the theories of stability and bifurcations, but also thermodynamic stability criteria indicate situations under which fluctuations play the fundamental role in the time evolution of water resources.

Zusammenfassung: In allen Zeitmaßstäben basiert die Entwicklung der Wasserressourcen auf Nichtlinearität, Instationarität und Schwellenprozessen. Deshalb muß die Theorie der Wasserressourcensysteme deterministische Methoden umfassen, um Prozesse entfernt von Bifurkationen zu behandeln, aber auch stochastische Methoden und Verfahren der Synergetik zur Behandlung der Dynamik in der Nähe der Bifurkationen der nichtlinearen deterministischen Differentialgleichungen. Nicht nur Stabilitäts- und Bifurkationstheorie, sondern auch thermodynamische Stabilitätskriterien zeigen Situationen an, in denen Fluktuationen die fundamentale Rolle in der zeitlichen Entwicklung von Wasserressourcen spielen.

Introduction

States and dynamics of water systems can mathematically be represented by "deterministic" models as well as by "stochastic" models. In both cases connections are established between the results of observations, on one hand, and the variables in the equations of the mathematical formalism on the other.

Usually, a deterministic model consists of differential equations with appropriate initial and boundary conditions. Deterministic models of water systems should be

[1]) Prof. Dr. rer. nat. habil. PETER MAUERSBERGER, Akademie der Wissenschaften der DDR, Institut für Geographie und Geoökologie, Bereich Hydrologie, Müggelseedamm 260, Berlin, DDR-1162.

based on the balances of mass, momentum, energy and entropy, taking into account laws governing chemical and biological processes in the water. These models sometimes are called 'physical' or 'phenomenological' or 'macroscopic' models, since they are founded on physical principles and introduce macroscopically measurable observables as dependent variables. Deterministic models are suited for the analysis of the major factors affecting processes, reveal the basic regularities inherent in the phenomenon, and permit, in general, prediction of the system's development. But under special conditions they cannot even describe certain phenomena which occur in natural systems.

Processes in nature are characterized by fluctuations often regarded as 'disturbances' or 'random deviations' from the mean values. If the investigator must not only identify the basic regularities, but also analyse possible deviations from them, stochastic methods or models, respectively, must play a decisive role. But the application of stochastic methods leads to mathematical difficulties and sometime is almost impossible.

The theory of water resources systems has to unite deterministic and stochastic methods in modelling efforts. Physical and stochastic methods should not be treated as rival approaches to water resources research, since in reality deterministic and stochastic elements are closely connected in the time evolution of water resources. We have to determine under which conditions deterministic methods are appropriate to deal with hydrological problems and under which conditions stochastic methods must be applied.

The very existence of many degrees of freedom in surface and subsurface water systems automatically implies the appearance of internally generated fluctuations of state variables and processes inside the system. In addition, fluctuations are introduced from outside. The macroscopic variables are subjects to deviations around certain "reference values". These deviations appear to the observer as fluctuations showing a frequency distribution which is typical for the state and process under consideration. Stable states are accompanied with small oscillations. The law of large numbers holds true. If the stability of a macroscopic state has been lost, by amplification of fluctuations the transition of the system to another state or even to another structure may happen. Especially in the vicinity of critical points or thresholds, there exist a number of instances where the deterministic description may not be adequate. The stochastic and the synergetic approach are suited to deal with situations when finite fluctuations are amplified, attain a macroscopic level, and drive the system to a new state (e.g. [1], [4], [11], [12], [13]).

Instabilities

In order to simplify the treatment of the problem, very often it is assumed that the system (or the process which is to be investigated) is linear, stationary, GAUSSIAN and non-threshold. But in all time-scales evolution is based on non-linearity, non-stationarity, non-GAUSSianity and on threshold processes or "phase-transitions". Equations describing natural phenomena, especially processes in water systems, are non-linear, time dependent, and involve threshold parameters, bifurcations, instabilities. Deviations from "normal" probability and/or density distributions occur. Under special conditions, the Law of Large Numbers breaks down. Since non-linear systems driven far from equilibrium can undergo instabilities and phase-transitions, it is necessary to apply synergetics and to construct a theory of fluctuations around far from equilibrium states as supplements to the predictions of the macroscopic description, especially in the neighbourhood of instabilities.

Far from instabilities

Far from instabilities of the macroscopic equations, the Law of Large Numbers is satisfied. The expectation values of a suitably chosen set of state variables correspond to the macroscopic variables used in the deterministic description. The stochastic treatment gives additional informations with respect to the macroscopic description: variances and other quantities reflecting explicitly the effects of fluctuations in non-linear systems. If the distribution is peaked around a well-defined value, the first moment, the mean, is directly related to the maximum of the probability distribution. Then both these values are expected to describe correctly the evolution of the macroscopic state. The average laws differ from the phenomenological ones, obeyed by the macroscopic variables, by terms related to the variances or to higher-order moments of the distribution of fluctuations. The central limit theorem imposes the order of magnitude of the variances of fluctuations relative to the mean value in such a way that the scales of the macroscopic averages and of the fluctuations are clearly separated. The relative importance of fluctuations diminishes as the size of the system increases:

$$\delta X_i/\overline{X} \sim n^{-1/2} \to 0 \quad \text{if} \quad n \to \infty \tag{1}$$

where δX_i and $\overline{X}$ denote fluctuation and mean, whereas n is related to the size of the system.

Thus, far from instabilities of the phenomenological equations the deterministic approach which does not take into account fluctuations explicitly allows to simplify or even to make possible the analysis and prognosis of processes in water systems.

In the vicinity of bifurcations

For special values of threshold parameters more than one physically acceptable solution (i.e. real and bounded solution) of the non-linear differential equations exist and very often the macroscopic equations do not justify preference to one of these solutions. Bifurcating solutions may even lose their stability properties. At the branching of at least two solutions the exchange of stability may take place. Then in the neighbourhood of those bifurcation points the system can evolve to a new regime exhibiting a new type of spatial and/or temporal order. A slight variation of the threshold parameter(s) results in qualitatively different trajectories in the phase space. The new solution emerging at the bifurcation point may depend on the threshold parameters in a non-linear fashion. General techniques dealing with such phenomena are

— stability theory (cf. LA SALLE, LEFSCHETZ [6], MINORSKY [12]),
— theory of bifurcations (cf. IOOS, JOSEPH [5], SALVADORA [14]),
— theory of catastrophes (cf. THOM [15]),
— synergetics (cf. HAKEN [3]).

Qualitative changes of structure and dynamics of non-equilibrium systems near bifurcation points are treated by NICOLIS, PRIGOGINE [13] and by EBELING, FEISTEL [2].

Instabilities far from equilibrium are situations when $\delta X/\overline{X}$ tends to zero more slowly than $n^{-1/2}$, remains constant or even tends to infinity as $n \to \infty$. The Law of Large Numbers breaks down. The clearcut distinction between the scales of macroscopic variables and fluctuations vanishes. Amplified fluctuations drive the average values to a new macroscopic state corresponding to another branch of the solution of the phenomenological equations, effecting phase-transitions, spontaneous formation of spatial compartments, generation of bursts of chemical and/or biological activity, and other self-organizing processes. If the probability distributions of state variables are multiply humped, we expect that the results of macroscopic measurements are related to the most probable values of the distributions and to the dispersion around them. Therefore, we are interested in estimating not only the a priori probability of having fluctuations of a certain size and range appearing within the system, but also in determining the growth rates of such fluctuations beyond instabilities. The onset of self-organizing processes implies that there exists a volume element in the system (of dimensions much larger than the characteristic molecular dimensions, but smaller than the volume of the total system) within which the fluctuations behave coherently, and that they add up to sufficient size and subsequently modify the macroscopic state. The variance of the fluctuations qualitatively changes as their scale increases. Properties as size, range of extent, and correlation length of fluctuations play an important role in the vicinity of bifurcations. Long-range correlations between macroscopic fluctuations emerge so that different macroscopic regions in the system do not evolve independently, but become coupled by these correlations. The description of fluctuations must necessarily be capable of differentiating between fluctuations of variable ranges and coherence lengths.

Synergetics

It is worth mentioning that methods of synergetics offer powerful tools to deal with fluctuations and to study the stochastic dynamics in the transitions near bifurcations (HAKEN [3]). Focussing attention to situations in which the macroscopic behaviour of the system undergoes drastic changes, synergetics demonstrates how cooperation of subsystems bring about spatial, temporal and/or functional structures in macroscopic scales. In cases when the system loses (linear) stability, a small number of state variables, the so-called "order parameters", sometimes only one state variable, become dominating in the sense that they govern all the other state variables ("slaving principle"). The complex of solving the full system of differential equations is reduced to the solution of a small system of equations or of one equation, respectively.

Thermodynamic stability criteria

There are several reasons for the application and combination of stochastic and deterministic methods in modelling and managing water resources systems. For instance, averaging procedures are applied to derive macroscopic balance equations for mass, momentum, energy and entropy from microscopic balances valid for very heterogeneous multiphase systems like groundwater resources. Furthermore, statistical handling of experimental data and field observations remains an important stage of the research process. But for management of water resources the investigation of 'critical situations' of aquatic ecosystems, when their regime becomes unstable and a new type of structure and functioning originates, is of at least the same importance as the simulation of the annual course of part of state variables usually being the aim of ecosystem modelling. Structural changes must be studied thoroughly before we can rely on forecasts with help of models, especially if external impacts strongly change. Progress in hydrology demands for better understanding of control mechanisms and strategies of aquatic ecosystems.

Therefore, we are interested in asking for conditions under which deterministic equations may be applied in modelling water systems and for situations when fluctuations play the fundamental role in the time evolution of water resources. The thermodynamic theory of aquatic ecosystems does not only support the unified deterministic treatment of physical, chemical and biological processes in surface and subsurface waters, but also indicates conditions under which the stochastic approach offers informations inevitably necessary in addition to the deterministic description.

The thermodynamic theory of aquatic ecosystems is based on the following 'macroscopic' equations (MAUERSBERGER [7], [8]):

— mass balances of all constituents (water, chemicals and biota),
— balance of momentum (NAVIER-STOKES equation),
— balance of kinetic, internal and electromagnetic energy,
— initial and boundary conditions,
— generalized GIBBS relation,
— entropy balance and Second Law of Thermodynamics,
— rates of biological processes as functions of state variables derived from an optimization principle (MAUERSBERGER [10]).

Deterministic models of water resources, derived from these basic equations, become non-linear differential equations. Such systems of differential equations can — apart from simple cases — be solved only by help of computers. It is a very difficult task or even impossible to get in all cases an overview of the set of possible solutions by numerical integration. Therefore, attention is called to the fact that the vanishing of a thermodynamic quantity, the excess entropy production, allows to determine bifurcation points [13].

The local entropy production σ, being a function of space and time, is given by

$$\sigma = \sum_{m} J_m X_m > 0 \tag{2}$$

where J_m denotes the 'fluxes' and X_m the 'forces' of the physical, chemical and biological processes in the water: diffusion, heat flow, friction, chemical reactions, hydrolysis, primary production, respiration, grazing, excretion, mortality etc. (cf. MAUERSBERGER [8], [9]). By the Second Law of Thermodynamics $\sigma > 0$, since all these processes are irreversible. The local excess entropy production $\delta^2\sigma$ is defined by

$$\delta^2\sigma = \sum_{m} \delta J_m \delta X_m , \tag{3}$$

arising from the excess flows $\delta J_m = J_m - J_m^*$ and excess forces $\delta X_m = X_m - X_m^*$, i.e. from the deviations of the state X_m, J_m from a stationary reference state X_m^*, J_m^*. The vanishing of $\delta^2\sigma$ or of $\int_{(V)} \delta^2\sigma \, \mathrm{d}V$, respectively, determines the thermodynamic threshold that separates the stable regime ($\delta^2\sigma > 0$) from the unstable regime ($\delta^2\sigma < 0$). Beyond the threshold another, perhaps radically different type of structure and functioning of the ecosystem comes into being. Without going into details, as a very simple example we consider primary production. If temperature, solar radiation, nutrients and oxygen concentration are assumed to remain constant, the driving force X and the flux J are proportional to the following functions of biomass B:

$$X \sim -\ln B, \qquad J \sim B\,[1 - (B/B_{\max})^n] . \tag{4}$$

Confining to variations δB of biomass, we get

$$\delta^2\sigma \sim [(1 + n)(B/B_{\max})^n - 1](\delta B)^2\,, \qquad (5)$$
$$\delta^2\sigma \gtreqless 0 \quad \text{for} \quad B \gtreqless B_{\max}/(1 + n)^{1/n}$$

and conclude that under certain conditions non-linear processes like (5) contribute to destabilization. Of course, further processes and dependences must be taken into account if the condition (3) is used to calculate constraints capable of inducing an instability. Furthermore, $B_{\max}$ depends on temperature, light intensity and on the local concentrations of dissolved oxygen and other chemicals taking part in primary production, so that $B_{\max}$ is a function of space and time. The quantity n depends on the deviation of the state of the biocoenosis from a stable stationary reference state ('Fließgleichgewicht'). Obviously the activities of the biocoenosis influence on the threshold values and stability.

Concluding remarks

Thermodynamic stability criteria can indicate situations under which transition processes in the time evolution of water resources result from amplified fluctuations, since entropy producing and entropy reducing processes inside the water system and across its boundary regulate the structure, state and further development of the system. Surface and subsurface water resources are open systems exchanging energy and matter with their surroundings. All the physical, chemical and biological processes in the water body are irreversible processes producing entropy. The flows of energy and matter through the water body and across its boundary effect the export of entropy produced inside. Therefore, the entropy balance equation can serve as a starting point for the investigation of stationary states and instabilities or bifurcations, respectively. It seems worth while further to improve these methods, since neither the hydrothermodynamic theory of water resources has been fully developed, nor are the possibilities of applying this theory exhausted.

References

[1] Arnold, L.; Lefever, R. (eds.): Stochastic nonlinear systems in physics, chemistry, and biology. — Proceed. Workshop, Bielefeld, Oct. 5—11. 1980. — Berlin, Heidelberg, New York: Springer-Verlag, 1981.

[2] Ebeling, W.; Feistel, R.: Physik der Selbstorganisation und Evolution. — Berlin: Akademie-Verlag, 1982.

[3] Haken, H.: Advanced synergetics. — Berlin, Heidelberg, New York, Tokyo: Springer-Verlag, 1983.

[4] HORSTHEMKE, W.; KONDEPUDI, D. K. (eds.): Fluctuations and sensitivity in nonequilibrium systems. — Proceed. Internat. Confer., Austin/Texas, March 12—16, 1984. — Berlin, Heidelberg, New York, Tokyo: Springer-Verlag, 1984.
[5] IOOS, G.; JOSEPH, D. D.: Elementary stability and bifurcation theory. — New York, Heidelberg, Berlin: Springer-Verlag, 1980.
[6] LA SALLE, J.; LEFSCHETZ, S.: Stability by LIAPUNOV's direct method with applications. — New York, London: Academic Press, 1961.
[7] MAUERSBERGER, P.: On the theoretical basis of modelling the quality of surface and subsurface waters. — Proceed. IIASA—IAHS Symposium Baden/Austria, Sept. 1978. — IAHS—AISH Publ. **125** (1978), 14—23.
[8] MAUERSBERGER, P.: General principles in deterministic water quality modeling. — In: ORLOB, G. T. (ed.): Mathematical modeling of water quality: streams, and reservoirs. — Chichester, New York, Brisbane, Toronto, Singapore: John Wiley & Sons, 1983, 42—115.
[9] MAUERSBERGER, P.: Local entropy production in aquatic ecosystems. — Acta hydrophys. **29** (1985) 4, 235—258.
[10] MAUERSBERGER, P.: Optimal control of biological processes in aquatic ecosystems. — Gerlands Beitr. Geophys. **94** (1985) 2, 141—147.
[11] MAY, R. M.: Thresholds and breakpoints in ecosystems with a multiplicity of stable states. — Nature **269** (1977), 471—477.
[12] MINORSKY, N.: Nonlinear oscillations. — Princetown, New York: Van Nostrand, 1962.
[13] NICOLIS, G.; PRIGOGINE, I.: Self-organization in nonequilibrium systems. — London, New York, Sydney, Toronto: John Wiley & Sons, 1977.
[14] SALVADORA, L. (ed.): Bifurcation theory and applications. — Lecture Notes in Mathem., Vol. 1057. — Berlin, Heidelberg, New York, Tokyo, 1984.
[15] THOM, R.: Stabilité structurelle et morphogénèse. — New York: Benjamin, 1972.

Acta hydrophys., Berlin 31 (1987) 3/4, S. 173–179

The Critical Depth of a Lake for the Maximum Release of Phosphorus from the Sediments of the Profundal Zone

By G. MOTHES †[1])

Summary: The critical depth of a lake is the depth with the highest sediment increase and release of phosphorus in the process of its ageing. It depends on the size of the lake and its exposure to wind, and probably also on the kind and the rate of sedimenting material. If this depth is known, it is possible to take optimal measures regarding the maintenance and improvement of water quality, nutrient elimination or humus production.

Zusammenfassung: Die kritische Seetiefe ist die Tiefe, in der der Sedimentzuwachs und die Phosphorfreisetzung im Alterungsprozeß des Sees am höchsten sind. Sie hängt von der Größe und von der Windexposition des Gewässers, wahrscheinlich auch von der Art und der Rate des sedimentierenden Materials ab. Ihre Kenntnis führt zur Optimierung von Bewirtschaftungsmaßnahmen zur Wasserqualität, Nährstoffelimination oder Humusgewinnung.

Резюме: Критической глубиной озера является такая его глубина, при прохождении которой в течении старения озера процессы растения седиментов и выпускания фосфора достигают свои количественные пики. Эта глубина зависит от размеров водоема и от экспозиции на ветер, наверно также от качества и количества седиментирующего материала. Знание этой величины может быть использовано для оптимизации мероприятий управления водоемов, как для улучшения качества воды, элиминации питательных веществ или добывания гумуса.

The release of phosphorus from profundal sediments is closely linked with the kind of decomposition of organic substance sedimenting there. If the hypolimnion is rich in oxygen, the decomposition preferably takes place with the formation of oxidation products such as CO_2. Simultaneously, phosphorus for the most part remains bound to particular substance [2]. However, if the hypolimnic amount of oxygen is not sufficient for the decomposition of organic substance in the profundal

[1]) Dr. rer. nat. habil. GEORG MOTHES, Akademie der Wissenschaften der DDR, Institut für Geographie und Geoökologie, Bereich Hydrologie, Müggelseedamm 260, Berlin, DDR-1162. — verst. am 27. 7. 1986.

zone and in the hypolimnion, there will be partial decomposition of organic substance in the sediment with the release of reduction products such as CH_4, H_2S and others from the sediment. The oxidation of these reduction products occurs in free water or in the air. Simultaneously with the release of reduction products from the sediment there will be a release of dissolved phosphate.

There is the following interrelation between morphometry and the hypolimnic oxygen level of a lake during the summer stagnation period [10]: the deeper a lake, the higher the hypolimnion which it has and the farther its position in the north (northern hemisphere), the more oligotrophic a lake will be (i.e. the higher its oxygen level in the hypolimnion during the summer will be). In the following we shall have a look at lakes situated in the north of the GDR, which means that the geographic aspect has to be ruled out. The integration of the lake in its catchment area will be considered later on.

The interrelation between the depth of the lake, the oxygen level in the hypolimnion and the release of phosphate from profundal sediments can be gathered from Table 1. It also suggests another interrelation between the decomposition of organic substance and the release of phosphate. If in the anoxic milieu the decomposition of organic substance is impeded and the release of phosphorus encouraged, the composition of the sedimented material changes at the point of transition to the permanent sediment. The atomic carbon — phosphorus ratio increases. Inversely, this ratio drops in case of oxidative decomposition of the sedimented substance if organic material is largely mineralized and phosphate remains bound to particles

TABLE 1
Characteristic data for the eutrophic lake Dagowsee and the oligotrophic lake Stechlinsee [5]

	Dagowsee	Stechlinsee
Maximum depth (in m)	9.5	68
Medium depth (in m)	5.0	22.8
Hypolimnic O_2 concentration in midsummer (in $mg \cdot l^{-1}$)	0	>5
P release from profundal sediment in midsummer (in $mg\ P \cdot m^{-2} \cdot d^{-1}$)	30	<1 (in most cases not detectable)
P supply from profundal sediment to the euphotic zone in midsummer (in $mg\ P \cdot m^{-2} \cdot d^{-1}$)	10	not detectable
Atomic C:P ratio		
— in sedimenting material	105	201
— in the sediment	425	107

in the sediment. This has been proved in surface sediments of different lakes [4]. Even though it is not possible to infer absolute P release rates from the C:P ratio, trends of changes of the C:P ratio should correlate with corresponding trends of P release rates in a lake.

The sediment increase in a lake results from the annual retention of phosphorus (import — export) and the concentration of phosphorus in the sediment. The concentration of phosphorus is closely connected with the ratio of organic carbon to phosphorus. Only chalk horizons in the region under analysis lead to a strong dilution of phosphorus in the sediment. This should be ignored in our comparison.

To explain the phenomenon 'critical depth of a lake' we assume the following ideal case: the lake is deep, bowlshaped and there is always the same P supply to the sediment ($mg\ P \cdot m^{-2} \cdot a^{-1} = const$). Initially, the sediment increase remains low owing to a high oxygen level in the hypolimnion and, consequently, a high P concentration in the sediment. Over thousands of years the hypolimnion grows smaller and smaller due to the accumulation of sediment, the oxygen level as well as the P concentration in the sediment drop. There is a rapid sediment increase up to the critical depth of the lake. The latter is to be defined, with a maximum duration of the summer stagnation and a minimum volume of the hypolimnion, as a depth where the anoxic phase of the sediment decomposition has a maximum extension. In this phase the highest C:P ratios are to be expected.

As the sediment continues to grow, the lake becomes shallower and the stagnation periods shorten. The periods of P release no longer last for months, but are reduced to weeks. The accumulation of reduction products in the sediment during the stagnation periods is counteracted by increasingly extended periods of oxidative decomposition in circulation periods. So the C:P ratio in the sediment will drop again, the P concentration will increase, and the sediment increase will decrease, too. The periodical release of P in a polymictic lake has been well analysed in the case of lake Großer Müggelsee [6], [8].

TABLE 2
Morphological data on shallow polymictic lakes [1], [3]

	Großer Müggelsee	Kleiner Müggelsee	Dämeritzsee
Surface (in km^2)	7.2	0.16	0.94
Volume (in $10^6\ m^3$)	35.3	0.47	2.68
Maximum depth (in m)	8	4.2	4.5
Medium depth (in m)	4.9	3.0	2.8

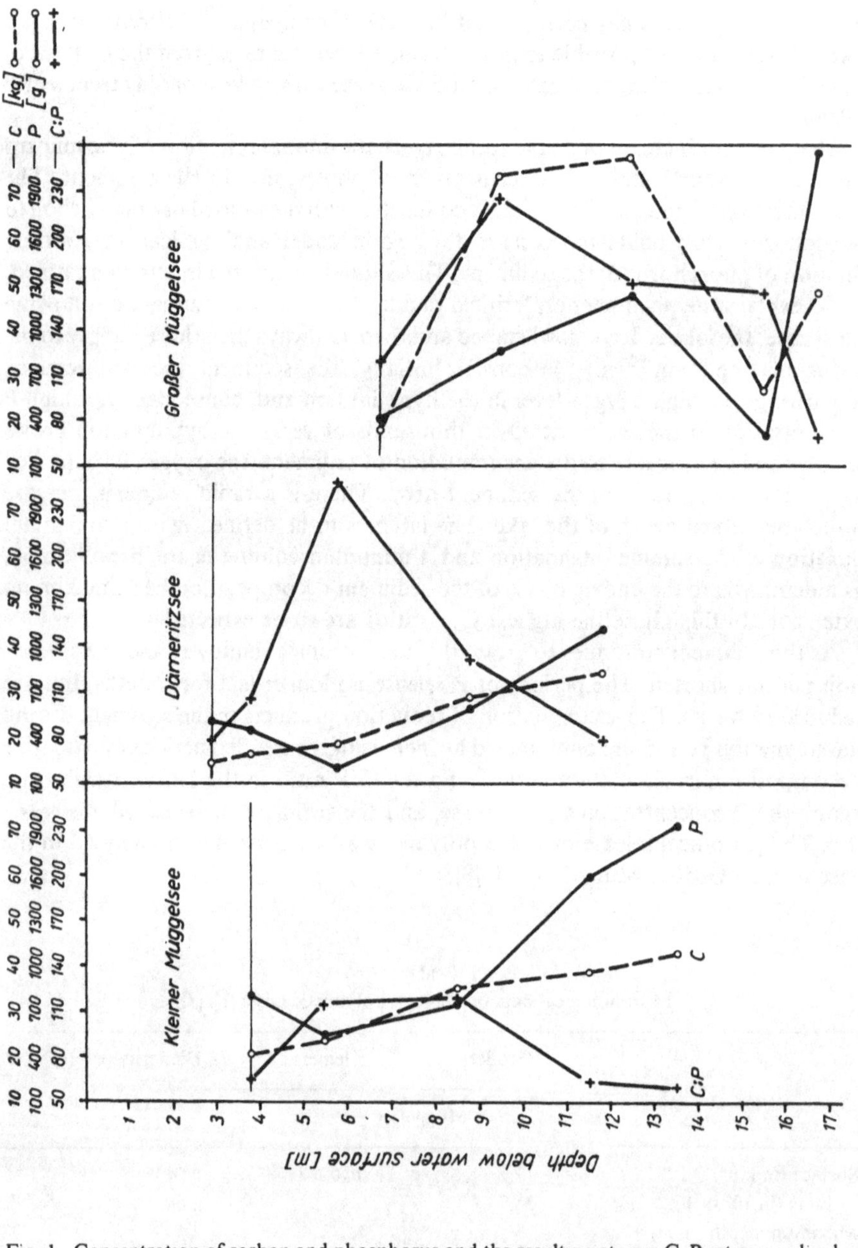

Fig. 1 Concentration of carbon and phosphorus and the resulting atomic C P ratio in individual samples taken from sediment profiles of different lakes

This ideal case does not take into account the fact that the phosphorus released from the sediment is anew absorbed by organisms and sedimentates again with carbon. This is the way the sediment increase is produced. What we have said suggests that the critical depth cannot be definitely observed in lakes with a stable stratification. It should become visible only in polymictic lakes if their sediments are perforated. The greater the lake and the more exposed it is to winds and the lower the specific rate of consumption of oxygen of the sedimenting material, the deeper its critical depth should be.

Perforation of the sediment by drilling carried out by 'ZBE Groß Kreuz' made it possible to analyse individual samples of sediment profiles from three lakes: Großer and Kleiner Müggelsee and Dämeritzsee. All these lakes are today relatively shallow (Table 2). The thickness of the sediment at the deepest points exceeds the high of the water column over it.

Fig. 1 contains extracts from the results of the analyses in their classification of depth, i. e. the absolute concentration of carbon and phosphorus as well as the atomic C:P ratio resulting from it. The absolute C and P concentrations increase with the depth of the sediment (exception: lake Großer Müggelsee, where there is a chalk horizon at a depth of 15 to 16 metres which has not been taken into account), a result of the increasing consolidation of the sediment due to the pressure of the sediment lying over it.

The C:P ratios increase from the bottom upward and drop again near the sediment surface. Only in lake Dämeritzsee the maximum C:P value seems to have been attained; in lake Kleiner Müggelsee it probably lies between two sample horizons. The estimated critical depths of the lakes are:

Kleiner Müggelsee: 7 m

Großer Müggelsee: 9 m (in a larger profile about 10 m)

Dämeritzsee: 6 m

The absolute value of the critical depth of a lake on the one hand depends on its size and its exposure to wind. Lake Dämeritzsee and Großer Müggelsee, which are both directly influenced by the river Spree, illustrate this (cf. Table 2). High exposure to wind strengthens the O_2 transport through circulation in greater depths. On the other hand, the specific consumption of oxygen and the absolute amount of sedimenting material seem to play a part — which points to the factor of the catchment area.

Lake Kleiner Müggelsee is the smallest of the three lakes but does not have the lowest critical depth. It is surrounded by a hilly east and south shore and is relatively sheltered from wind. The dense tree populations on its shores form an additional shield from the wind. All that should cause a low critical depth of the lake. However, the Spree does not flow through the lake. The sediment profile has been taken from the back basin, which is isolated from the reaches of the Spree. Possibly, the sedimentation of fallen leaves leads here to an important accumulation of sediment.

Thus the sediment in the depth of lake Kleiner Müggelsee is obviously richer in C (over 20 percent of the dry matter) than that of lake Großer Müggelsee (below 20 per cent of the dry matter). This allochthonous material, which is added during the autumn circulation, hardly consumes any oxygen during the summer stagnation period in the following year, but binds phosphorus. Given a low oxygen consumption of the sedimented material, small oxygen supplies to the sediment can possibly influence the kind of sediment composition to a decisive degree due to an unstable thermocline. These problems, however, need to be analysed.

The critical depth of a lake suggests that at this depth there is a maximum sediment increase (lowest specific P concentration in relation to C). An artificial depression of the body of water in this sector is most rapidly levelled out. The fact that in this sector a maximum release of P from the sediment occurs can only be inferred from the previous observations.

The theoretical interrelations regarding this critical depth of a lake were described by RYDING [9]. Initial results of measurements were given by MOTHES [7]. If measures are taken to change the depth of a lake, the following observations regarding the critical depth should be taken into account:

1. If a lake is made deeper to reduce the nutrient export from the sediment to the euphotic zone, one should go beyond the critical depth.
2. The previous results of analyses of the phosphate budget in lakes suggest that in this sector there is the highest P release from the sediment and the highest P supply to the euphotic zone through internal loading. If a shallow lake is deepened until the critical depth is reached, the quality of water will worsen.
3. In case of a permanent extraction of sediment in order to produce humus and plant fertilizer, the critical depth of a lake should be preferred, for it is there that the highest accumulation of organic detritus occurs in the permanent sediment in relation to the constant accumulation of nutrients in large sections.
4. The critical depth of a lake is a parameter specific to a body of water, which also depends on the kind of the catchment area. It cannot yet be calculated and must be determined by experiment. But it should be of importance for calculating nutrient elimination, particularly in bodies of water with a strong flow (e.g. pre-impoundments). Moreover, it will make it possible to determine measures for the maintenance of equipment for the elimination of nutrients.

References

[1] DRIESCHER, E.; TITTMANN, M.; MOTHES, G.: Beschreibung des Untersuchungsgebietes Müggelsee. — Acta hydrophys. **26** (1981), 183—188.

[2] EINSELE, W.: Über die Beziehung des Eisenkreislaufs zum Phosphatkreislauf im eutrophen See. — Arch. Hydrobiol. **29** (1936), 653—677.

[3] EYRICH, A.: Limnologische Untersuchungen in Berliner Gewässern zur Ermittlung der Verbreitung rezenter und subrezenter Sedimente und deren Bedeutung für die Uferfiltration. — Berlin: Humboldt-Universität, Dissertation, 1980, 99 S.

[4] MOTHES, G.: Die Nährstoffverteilung in Sedimenten norddeutscher Seen und Hinweise auf den Austausch Sediment — Pelagial. — Acta hydrochim. hydrobiol. **8** (1980), 35—46.

[5] MOTHES, G.: Sedimentation und Stoffbilanzen in Seen des Stechlinseegebietes. — Limnologica **13** (1981), 147—194.

[6] MOTHES, G.: Sauerstoff- und Phosphorgehalte im Müggelsee als Indikatoren für Austauschprozesse. — Acta hydrophys. **27** (1982), 347—356.

[7] MOTHES, G.: Die organischen Sedimente und Binnenseen. — Wiss. Fortschr. **35** (1985), 27—28.

[8] MOTHES, G.; DOLLAN, L.; HOFFMANN, I.; KOZERSKI, H.-P.; STEINMANN, J.: Der Phosphoraustausch zwischen Sediment und Pelagial im Großen Müggelsee. — Acta hydrophys. **30** (1986), 91—136.

[9] RYDING, S.-O.: Chemical and microbiological processes as regulators of the exchange of substances between sediments and water in shallow eutrophic lakes. — Int. Rev. ges. Hydrobiol. **70** (1985), 657—702.

[10] THIENEMANN, A.: Die Binnengewässer Mitteleuropas. Eine limnologische Einführung. — Binnengewässer **1** (1925), 1—255.

Acta hydrophys., Berlin 31 (1987) 3/4, S. 181–190

Long-Term Variations of Primary Production in an Eutrophic Shallow Lake with Special Regard to Photosynthetic Efficiency of Blue-Green Algae

By B. NIXDORF and R. STELLMACHER[1])

Summary Statistical treatment of measuring series of primary production and biomass of phytoplankton and temperature as well as comparison of derived biological parameters for the investigation period 1978–1984 resulted in a decreasing trend of biological dates despite favourable meteorological conditions. Photosynthetic efficiency of phytoplankton decreased discontinuously with the development of colony forming blue-green algae Reasons for this phenomenon are supposed in the wide range of metabolic possibilities of cyanophyta, especially for colony forming species, to overcome extreme situations as to nutrients or irradiance. Several strategies are discussed.

Zusammenfassung Die statistische Bearbeitung von Meßreihen der Primärproduktion und der Biomasse des Phytoplanktons und der Wassertemperatur sowie die Gegenüberstellung daraus abgeleiteter biologischer Parameter für die Untersuchungsjahre 1978–1984 ergaben eine fallende Tendenz der untersuchten biologischen Daten trotz günstiger meteorologischer Bedingungen. Die photosynthetische Effektivität des Phytoplanktons erniedrigte sich sprunghaft mit dem Aufkommen kolonienbildender Blaualgen Diese Erscheinung wird anhand verschiedener Strategien im Stoffhaushalt von Blaualgen, insbesondere von koloniebildenden, als Anpassungsreaktion auf Stress-Situationen diskutiert.

Резюме. Статистический анализ измерительных рядов первичной продукции и биомассы фитопланктона в исследенном периоде 1978—1984 показывали убывающий тренд в озере Мюггельзее Метеорологические условия были благоприятны. Фотосинтетическая эффективность фитопланктона унижалась скачкообразно с появлением колоний формирующих синезелёных водорослей. Это явление обсуждается на основании разных стратегий проведения синезелёных водорослей в частности при ситуциях стресса.

[1]) Dr rer nat. BRIGITTE NIXDORF, Dr sc. nat. RITA STELLMACHER, Akademie der Wissenschaften der DDR, Institut für Geographie und Geoökologie, Bereich Hydrologie, Müggelseedamm 260, Berlin, DDR-1162.

Since 1978 systematic investigations have been carried out for measuring biomass and primary production of phytoplankton ([13], [27]) in lake Müggelsee, an eutrophic shallow lake. These investigations were part of a complex programme dealing with circulation of matter and energy flow in the lake. Samples were collected and incubated with $NaH^{14}CO_3$ in situ at intervals of one (March—October) or, respectively, two (November—February) weeks. Accordingly, simultaneous measuring series are at disposal for the period 1978—84. For their statistical

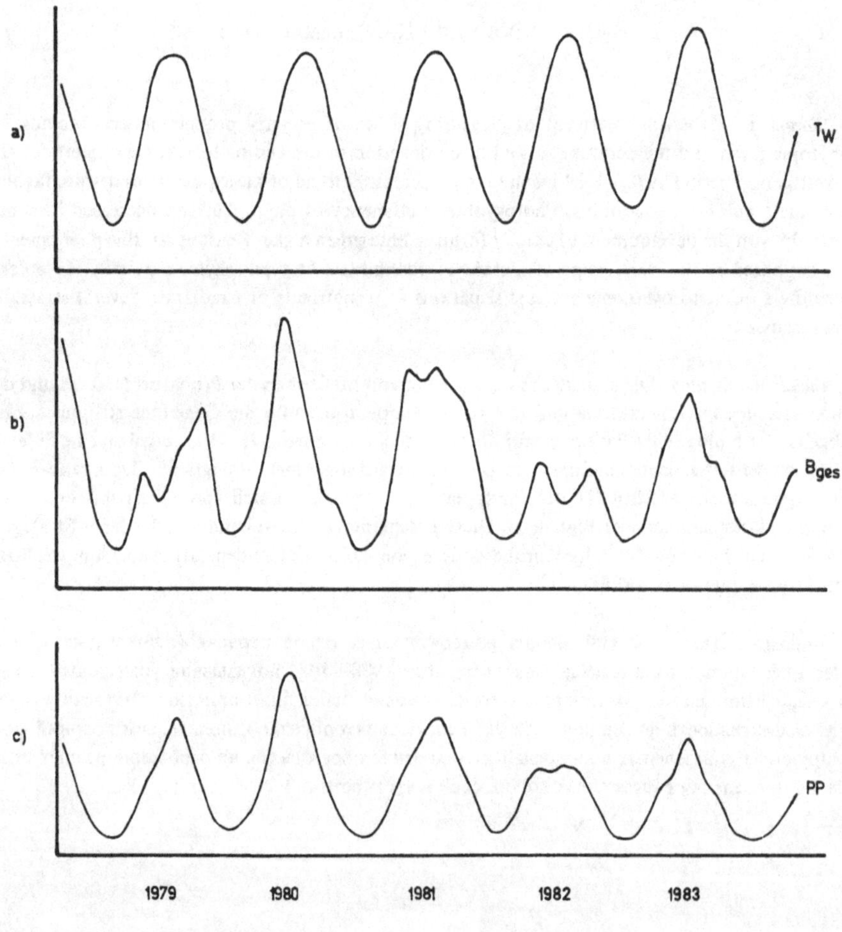

Fig. 1 Time series (1978—1984) smoothed with local spline-function
a) water temperature T_W (in °C), b) biomass B_{ges} (in g freshweight · m^{-3}), c) primary production PP (in g C · m^{-2} d^{-1})

treatment a measuring interval of 14 days was taken as a basis. Besides time behaviour of primary production and biomass also the water temperature was considered in the statistical analysis of time series.

To make the characteristics of time series more evident, the measuring numbers were smoothed (Fig. 1). The number of used weights determines the shortening of smoothed series correspondingly. Well-approved methods of modern time series analysis were applied to estimate the structure of a given time series. This analysis is based on a simple additive model. Trend tests were carried out by the nonparametric test according to MANN-KENNDALL. The probable start of trends was found out by help of progressive analysis. A description of the methods incl. examples of application is found in ([45], [47]). For finding periodicities, the method of maximum entropy spectral estimation was used, which had proved to be favourable for short-time series. OLBERG, RÁKÓCZI ([28]) describe the algorithm incl. FORTRAN programmes for calculation of spectra.

The statistical analyses gave the following results, taking a level of significance probability of 95%:

The time series of primary production exhibits a decreasing trend, what is significant for the smoothed series, but the total series does not fall below the critical value. It cannot be decided if it is a question of a partly realized wave in this case. Elimination of the trend results an estimation of the spectrum by a dominant

TABLE 1
Average values of biological and meteorological parameters for vegetation period (March/April—October) in the investigation periods 1978—1984

	1978	1979	1980	1981	1982	1983	1984
Irradiance (in $MJ \cdot m^{-2} \cdot d^{-1}$)	13.32	14.07	13.69	13.46	15.28	14.39	12.77
Temperature (in °C)	14.63	15.06	14.26	15.69	16.35	16.26	14.87
Primary production (in $g\,C \cdot m^{-2} \cdot a^{-1}$)	639	420	588	505	520	421	370
Biomass of algae (in g freshweight $\cdot m^{-3}$)	24.29	10.74	13.49	15.19	9.84	11.87	12.92
Biomasse of Cyanophyta (in g freshweight $\cdot m^{-3}$)	15.05	2.16	5.38	4.99	5.25	4.57	4.08
Biomass of diatoms (in g freshweight $\cdot m^{-3}$)	7.55	7.99	7.51	9.20	4.48	5.24	6.36
Photosynthetic efficiency $= \frac{\text{primary production}^{1)}}{\text{biomass}^{2)} \times \text{irradiance}}$	1.77	2.47	2.75	2.47	1.85	1.43	1.56

[1]) in $kJ \cdot m^{-2} \cdot d^{-1}$; 1 gC ≙ 40 kJ

[2]) in $kJ \cdot m^{-3}$; 1 g freshweight ≙ 3,6 kJ

yearly wave. The residual series, too, which belongs to the smoothed series, possesses a yearly wave because of the seasonal dependence of variability of the measured values. The series of total biomass shows a structure similar to that of primary production, however, the trend is not so clear, remaining below the level of significance.

The temperature series is determined by the annual variation. The values tend to slightly increase on an average, but this trend is not significant either.

Table 1 is to complete Fig. 1 and the results of statistical analysis. Values of yearly or seasonal averages of primary production and important parameters influencing primary productivity are compiled in it. Lake Müggelsee belongs to the eutrophic-hypertrophic aquatic ecosystems because of its relatively high primary production ([44]).

The biological values in Table 1 illustrate the retrograde trend demonstrated already by time series analysis. As a central problem the following question was put: What are the reasons for decrease in primary production in lake Müggelsee despites the rather increasing trends of factors favouring production such as temperature and light? Within the frame of this paper we concentrate mainly on the primary product itself — the biomass of phytoplankton — and its variations in the investigated period, because neither meteorological parameters nor the simultaneously estimated concentrations of nutrients in the lake showed obvious changes or trends in the period 1978—84.

Analysis of time series of biomass resulted in a similar, but less distinctly marked trend compared to primary production. Although in 1983 and 1984 a slight increase in phytoplankton biomass was observed in comparison with the preceding years (except for 1978), just in these two years minima of primary production were measured. Coincidentally, the following succession within the species structure could be observed with regard to Cyanophyta: Before 1981 colony forming blue-green algae appeared in negligible amounts beside the dominant species *Oscillatoria agardhii* and *O. redekei*. But since 1981/82 the colony forming species *Aphanizomenon flos-aquae* and *Microcystis aeruginosa* displace the *Oscillatoria* in summer and maintain a relatively high biomass frequently until autumn.

For characterization of photosynthetic efficiency we introduce here, in connection with ROTT [40] and TILZER [50], a parameter called "inner energy efficiency". It denotes the relation between primary production on one hand and biomass and light on the other. From Table 1 it is obvious that just since 1982 the increased occurrence of colony forming blue-green algae was connected with a rapid decrease of photosynthetic efficiency.

In what follows, some essential particularities are emphasized as possible reasons for the favoured development of blue-green algae with special regard to colony forming. Therefrom, some hypotheses for the decreased photosynthetic efficiency of these algae are to be derived.

1. Regulation of buoyancy

Most of the blue-green algae possess gas vacuoles and are therefore able to regulate their buoyancy by changing of turgor pressure. Thus, these organisms become positioned at that point within some physicochemical gradient which is most favourable for their growth ([8], [21], [22], [29], [37]).

2. Ability of CO_2-uptake from the atmosphere

Under certain conditions algae do not stratify but float to the water surface and form a water bloom, notwithstanding the high light intensity there. This phenomenon appears to have an ecological function, giving Cyanophyta access to atmospheric CO_2 when the concentration of inorganic carbon is growth limiting in the water ([1], [18], [19], [32], [33], [38], [49]).

Unfortunately, systematic investigations of free CO_2 have been carried out in lake Müggelsee as late as since 1982 with the following results supporting the above hypothesis: Free CO_2 is mostly not detectable in the lake pelagial when *Aphanizomenon* and *Microcystis* are the dominant species among the blue-green algae. Paerl [34] could demonstrate by autoradiography that only the external layers of colonies fix CO_2, leading to a distinct decrease of photosynthetic efficiency.

3. N-fixation

Some blue-green algae, as the only algae generally speaking, are able to overcome deficiency of inorganic nitrogen in water through fixation of atmospheric nitrogen. It was evidenced by measurements in lake Müggelsee that *Aphanizomenon* and, to a smaller extent, *Anabaena* fix nitrogen from the atmosphere when only minimum concentrations of inorganic nitrogen were detectable in the water ([3]). This kind of N-uptake causes a considerable decrease of photosynthetic efficiency ([20]). Thus, in 1979 a decrease of 50—80% was measured during a short-time development of *Aphanizomenon* in lake Müggelsee as a result of N-fixation.

4. Particularities of phosphate uptake

Cyanophyta occur also when phosphate ought to be a primary production limiting factor. However, it was shown that blue-green algae possess a high storage capacity for phosphate ([7], [23]), which is stored in the cells as polyphosphate. Moreover, by vertical migration close to the sediment of the lake the possibility for Cyanophyta is given to fill up the phosphate pools in stagnation periods ([17]). Some species are even able to hydrolyse organic phosphorous compounds ([25]).

5. Reactions of Cyanophyta at formation of nutrient poor microlayers

The supply of nutrients will deteriorate around single cells when there is a high primary production but no possibility to fill up nutrient depleted microlayers by turbulent mixing of the water body. In this way, primary production of phytoplankton becomes less effective ([46]), however, just blue-green algae are ecologically favoured in such situations because they have better possibilities than e.g. diatoms to take up nutrients from larger distances ([30]).

6. Composition of photosynthetic pigments

Beside chlorophyll, blue-green algae possess phycocyanin and phycoerythrin as accessory pigments. The absorption maxima for these pigments are in that range of visible radiation in which other algae cannot absorb energy. Moreover, some Cyanophyta produce carotenoides protecting chlorophyll from photooxidation at high intensity ([33]).

7. Protection from high light intensity and grazing pressure by colony forming

If colony forming blue-green algae float to the water surface and form a water bloom despite high light intensity, this results, among other things, in a displacement of the production maximum from depths of about 0.5 m to the surface, were other algal groups show a distinct inhibition of primary production ([48]). By agglomerating of algal cells and filaments some blue-green algae protect themselves from intensive solar radiation ([16]); however, this will enormously deteriorate the energy supply inside the colonies, i.e. photosynthetic efficiency decreases.

On the other hand, growing of colonies increases the euphotic layer and thus improves light transmission of the water body ([14], [39]).

Moreover, larger conglomerates of colonies diminish the risk to be grazed by zooplankton ([14], [51]).

8. Associations with bacteria — high decomposition rate — rapid recycling of nutrients

Cyanophyta are known to be closely associated to bacteria ([10], [15], [24], [26], [31]). This tight coupling can stimulate algal growth by a rapid bacterial remineralisation of algal metabolic products such as exudates or autolysis products. As proposed by FALLON and BROCK ([5], [6]), the destruction rate of Cyanophyta is higher compared to other algal groups. This can improve the pelagic nutrient situation ([35]). We obtained similar results by investigating decomposition by means of radioactively marked algae in lake Müggelsee.

9. Allelopathy

Cyanophyta can produce toxic metabolic compounds ([9]) and in this way suppress the development of other algal groups ([17], [52]) or zooplankton.

10. Heterotrophy

Another ecological advantage results from the ability of several blue-green algae to meet their requirements of energy by heterotrophic way of metabolism ([11], [12], [41], [42], [49]).

11. Wintering of blue-green algae on sediment surface

Mainly for the blue-green alga **Microcystis** it was shown that it can maintain its photosynthetic capability by wintering on the sediment surface. Therefore, this alga yields an inoculum of colonies for the development of the epilimnetic population in the subsequent summer ([2], [4], [36], [49]).

There are therefore a number of possibilities for blue-green algae, especially for colony forming species, to be competitive in eutrophic lakes during extreme situations as to nutrients or irradiance. In most cases these strategies of adaption, however, lead to a loss of photosynthetic efficiency. This could be demonstrated distinctly during N-fixation also in lake Müggelsee. Another reason for the decreased photosynthetic efficiency in lake Müggelsee may be that within the diatom structure *Stephanodiscus* became the dominant species which continuously increased its cell volume in the investigation period and also in seasonal cycle ([43]).

References

[1] BOOKER, M. J.; WALSBY, A. E.: Bloom forming and stratification by a planktonic blue-green alga in an experimental water column. — Br. Phycol. J. **16** (1981) 4, 411—422.

[2] CACERES, O.; REYNOLDS, C. S.: Some effects of artificially-enhanced anoxia on the growth of *Microcystis aeruginosa* KÜTZ. emend. ELENKEN, with special reference to the initiation of its annual growth cycle in lakes. — Arch. Hydrobiol. **99** (1984) 3, 379—397.

[3] DUDEL, G.; NICKLISCH, A.; KOH, J.-G.: Zum Einfluß von Licht und Temperatur auf die spezifische Wachstumsrate der Stickstoff-fixierenden Blaualge *Aphanizomenon flos-aquae* RALFS ex BORN et FLAH. (Cyanophyta) unter Berücksichtigung der Form der Stickstoffernährung. — Forschungsbericht, 1981 (unveröff.).

[4] FALLON, R. D.; BROCK, T. D.: Decomposition of blue-green algal (cyanobacterial) blooms in lake Mendota, Wisconsin. — Appl. Environm. Microbiol. **37** (1979) 5, 820—830.

[5] FALLON, R. D.; BROCK, T. D.: Planktonic blue-green algae: Production, sedimentation, and decomposition in lake Mendota, Wisconsin. — Limnol. Oceanogr. **25** (1980) 1, 72—88.

[6] FALLON, R. D.; BROCK, T. D.: Overwintering of *Microcytis* in lake Mendota. — Freshw. Biol. **11** (1981), 217—226.

[7] GEORGIEVA, V.: Effect of phosphorus nutrition on some aspects of *Microcystis* vitality under conditions existing in natural water basins. (russ.) — Fiziol. Rast. (Sofia) **7** (1981) 3, 52—56.

[8] GRANT, N. G.; WALSBY, A. E.: The contribution of photosynthate to turgor pressure rise in the blue-green alga *Anabaena flos-aquae*. — J. exp. Bot. **28** (1977), 409—415.

[9] HENNING, M.; KOHL, J.-G.: Toxische Blaualgen und ihre Toxine. — Biol. Rdsch. **19** (1981), 217—229.

[10] HERBST, V.; OVERBECK, J.: Metabolic coupling between the alga *Oscillatoria redeckei* and accompanying bacteria. — Naturwissenschaften **65** (1978), 598—599.

[11] HOARE, D. S.: Heterotrophic potentialities of the blue-green algae. — In: DESIKACHARY, T.V. (Ed.): Taxonomy and biology of blue-green algae. — Madras: 1972, 501—507.

[12] HOARE, D. S.; HOARE, S. L.; MOORE, R. B.: The photoassimilation of organic compounds by autotrophic blue-green algae. — J. Gen. Microbiol. **49** (1977), 351—370.

[13] HOEG, S.: Betrachtungen zur Jahresdynamik des Photoplanktons im Müggelsee. — Acta hydrophys. **28** (1983) 1/2, 5—36.

[14] HOLM, N. P.; GANF, G. G.; SHAPIRO, J.: Feeding and assimilation rates of *Daphina pulex* fed *Aphanizomenon flos-aquae*. — Limnol. Oceanogr. **28** (1983) 4, 677—687.

[15] HOPPE, H.-G.: Blue-green algae agglomeration in surface water: a microbiotope of high bacterial activity. — Kieler Meeresforsch., Sonderh. **5** (1981), 291—303.

[16] HORSTMANN, U.: Eutrophication and mass development of blue-green algae in the Baltic-Merentutkimuslait. Julk./Havsforskningsinst. Skr. No. 239 (1975), 83—90.

[17] KEATING, K. J.: Blue-green algal inhibition of diatom growth: transition from mesotrophic to eutrophic community structure. — Science **199** (1978), 971—973.

[18] KLEMER, A. R.; FEUILLADE, J.; FEUILLADE, M.: Cyanobacterial blooms: nitrogen limitation have opposite effects on the buoyancy of Oscillatoria. — Science **215** (1980), 4540, 1629—1631.

[19] KLEMER, A. R.; PIERSON, D. C.; WHITESIDE, M. C.: Blue-green algal (cyanobacteria) nutrition, buoyancy and bloom formation. — Verh. Internat. Verein. Limnol. **22** (1985) 5, 2791—2798.

[20] KOHL, J.-G.; BAIEROVA, J.; DUDEL, G.: Die Bedeutung der stickstofffixierenden Blaualgen für den Stoffhaushalt stehender und gestauter Binnengewässer. — Acta hydrochim. hydrobiol. **10** (1982) 5, 415—437.

[21] KONOPKA, A.: Buoyancy regulation and vertical migration by *Oscillatoria rubescens* in Crooked Lake, Indiana (USA). — Br. Phycol. **17** (1982) 4, 427—442.

[22] KONOPKA, A.; BROCK, T. D.; WALSBY, A. E.: Buoyancy regulation by planktonic blue-green algae in lake Mendota, Wisconsin. — Arch. Hydrobiol. **83** (1978) 4, 524—537.

[23] LANG, D. S.; BROWN, E. J.: Phosphorus-limited growth of a green algae and a blue-green alga. — Appl. Environ. Microbiol. **42** (1981), 1002—1009.

[24] LANGE, W.: Enhancement of algal growth in cyanophyta bacteria systems by carbanaceous compounds. — Can. J. Microbiol. **17** (1971), 303—314.

[25] LIVINGSTONE, D.; WHITTON, B. A.: Water chemistry and phosphatase activity of the blue-green alga *Rivularia* in upper teesdale sreams. — J. Ecol. **72** (1984) 2, 405—421.

[26] MEFFERT, M. E.; OVERBECK, J.: Interactions between *Oscillatoria redekei* (Cyanophyta) and bacteria. — Verh. Internat. Verein. Limnol. **21** (1981), 1432—1435.

[27] NIXDORF, B.: Photosynthetic activity of dominant algal species in eutrophic shallow lake (Großer Müggelsee, Berlin) investigated by microautoradiography. — Int. Revue ges. Hydrobiol. **70** (1985) 2, 237—246.

[28] OLBERG, M.; RÁKÓCZI, F.: Informationstheorie in Meteorologie und Geophysik. — Berlin: Akademie-Verlag, 1984.

[29] OLIVER, R. L.; WALSBY, A. E.: Direct evidence for the role of light-mediated gas vesicle collapse in the buoyancy regulation of *Anabaena flos-aquae* (Cyanobacteria). — Limnol. Oceanogr. **29** (1984) 4, 879—886.

[30] OLOFSSON, J. A.: The role of microlayers in controlling phytoplankton productivity. — In: Developments in Hydrobiology, Vol. 2, Hypertrophic Ecosystems, ed. by BARICA, J.; MUR, L. R. — The Hague, Boston, London: 1980, 83—93.

[31] PAERL, H. W.: Specific association of the blue-green algae *Anabaena* and *Aphanizomenon* with bacteria in freshwater blooms. — J. Phycol. **12** (1976) 4, 431—435.

[32] PAERL, H. W.; USTACH, J. F.: Blue-green algal scums: an explanation for their occurrence during freshwater blooms. — Limnol Oceanogr. **27** (1982) 2, 212—217.

[33] PAERL, H. W.; TUCKER, J.; BLAND, P. T.: Carotenoid enhancement and its role in maintaining blue-green algal (*Microcystis aeruginosa*) surface blooms. — Limnol. Oceanogr. **28** (1983) 5, 847—857.

[34] PAERL, H. W.: Partioning of carbon dioxide fixation in the colonial cyanobacterium *Microcystis aeruginosa*: Mechanism promoting formation of surface scums. — Appl. Environ. Microbiol. **46** (1983) 1, 252—259.

[35] PHILIPOSE, M. T.: Notes on the ecology, role and possible control of some common blue-green algal blooms in indian freshwater fish ponds. — In: DESIKACHARY, T. V. (Ed.): Taxonomy and biology of blue-green Algae. — Madras: 1972, 455—460.

[36] PRESTON, T.; STEWART, W. D. P.; REYNOLDS, C. S.: Bloom-forming cyanobacterium *Microcystis aeruginosa* overwinters on sediment surface. — Nature **288** (1980), 365—366.

[37] REYNOLDS, C. S.: Interrelation of photosynthetic behaviour and buoyancy regulation in natural population of a blue-green alga. — Freshwat. Biol. **5** (1975), 323—338.

[38] RIJN, VAN J.; SHILO, M.: Buoyancy regulation in a natural population of *Oscillatoria* spp. in fishponds. — Limnol. Oceanogr. **28** (1983) 5, 1034—1037.

[39] ROBARTS, R. D.; ZOHARY, T.: *Microcystis aeruginosa* and underwater light attenuation in a hypertrophic lake (Hartbeespoort Dam, South Africa). — J. Ecol. **72** (1984) 3, 1001—1017.

[40] ROTT, E.: Primary productivity and activity coefficients of the phytoplankton of a mesotrophic softwater lake (Pitburgersee, Tirol, Austria). — Int. Revue ges. Hydrobiol. **66** (1981) 1, 1—27.

[41] SAHU, J.: Heterotrophic growth and pigment composition of four filamentous blue-green algae. — Arch. Hydrobiol. **63** (1982), 189—199.

[42] SAUNDERS, G W Potential heterotrophy in a natural population of *Oscillatoria agardhii* var. *isothrix* SKUJA. – Limnol. Oceanogr **17** (1972), 704–711.

[43] SCHELLENBERGER, G. ; HOEG, S. Seston, Phytoplankton und Wasserfarbe in zwei verbundenen Flachsen. – Acta hydrophys. **30** (1986) 4, 231–273.

[44] Nutzung und Schutz der Gewässer, Klassifizierung der stehenden Gewässer (Seestandard). – Fachbereich-Standard, TGL 27885/01, Gruppe 11800, Entwurf 1981.

[45] SNEYERS, R. Sur l'analyse statistique des series d'observations. – Genève WMO TN 143, 1975.

[46] SREENIVASAN, A. Ecology of blue-green algae in tropical inland waters. – In DESIKACHARY, T V (Ed.). Taxonomy and biology of blue-green algae. – Madras 1972, 461–469.

[47] STELLMACHER, R. Prüfung der Homogenität und statistischen Analyse von Zeitreihen, vorgestellt am Beispiel klimatologischer und hydrologischer Reihen. – Berlin AdW der DDR, Diss. B, 1983.

[48] STELLMACHER, R.; NIXDORF B. Ein Beitrag zur Anwendung der Regressionsanalyse im Zusammenhang mit der Schätzung der Primärproduktion im Großen Müggelsee (Berlin) für den Zeitraum 1979–1982. – Acta hydrochim. et hydrobiol. **12** (1984) 6, 653–657

[49] STEWART, W D P Algal physiology and biochemistry – Bot. Monogr Vol. 10. – Oxford, London, Edinbourgh, Melbourne, Blackwell Scientific Publications, 1974, 987 S.

[50] TILZER, M. Dynamik und Produktivität von Phytoplankton und pelagischen Bakterien in einem Hochgebirgssee (Vorderer Finstaler See, Österreich). – Arch. Hydrobiol., Supp. 40 (1972), 201–273.

[51] UEHLINGER, U Zur Ökologie der planktischen Blaualge *Aphanizomenon flos-aquae* in einem Alpenrandsee. – Schweiz Z. Hydrol. **43** (1981) 1, 69–88.

[52] VINCENT, W F ; SILVESTER, W B. Growth of blue-green algae in the Manukon (New Zealand) oxidation ponds. – Water Res. **13** (1979) 8, 711–723

Acta hydrophys., Berlin **31** (1987) 3/4, S. 191–205

Soil Moisture and Runoff Components from a Physically Founded Approach

By G. PESCHKE[1])

Summary A simple, physically founded algorithm is depicted that makes it possible to calculate the increase in soil moisture and the resulting runoff components called overland flow and interflow which occur after rain events. This procedure can be coupled with submodels of percolation and evapotranspiration so as to make possible a longer-term dynamic simulation of the soil water balance.

Zusammenfassung Es wird ein einfacher, bodenphysikalisch begründeter Algorithmus dargestellt, der es gestattet, die Erhöhung der Bodenfeuchte und die entstehenden Abflußkomponenten Oberflächenabfluß und hypodermischer Abfluß als Folge von Regenereignissen zu berechnen. Dieses Verfahren kann mit Teilmodellen der Versickerung und der Evapotranspiration gekoppelt werden, um die längerfristige, dynamische Simulation des Bodenwasserhaushaltes zu ermöglichen.

1. Introduction

A number of application of infiltration models require both information concerning the moisture dynamics in the soil profile and the further division of infiltrated water into soil moisture storage, interflow and ground water recharge. For this purpose the description of infiltration as interfacial process based on empirical equations is generally insufficient. Instead, a physically founded approach is needed which, in the interest of an application to major systems, is simple enough and also reflects the rain intensity and the antecedent soil moisture as major quantities influencing the infiltration process.

A model that meets these requirements has been derived by the autor through the consistent pursuance of international efforts ([1], [2], [4], [5], [6], [13]) for the

[1]) Prof Dr sc. nat. GERD PESCHKE, Technische Universität Dresden, Sektion Wasserwesen, Bereich Hydrologie und Meteorologie, Mommsenstraße 13, Dresden, DDR-8027

following cases: constant precipitation intensity and layered soil profile ([7], [8], [9]) as well as time-variable intensity of rainfall and homogeneous soil ([10], [11]). Although these versions provided important findings and permitted useful applications, it was desirable from the scientific point of view and necessary in terms of providing an unswer to practical questions, to consider the generalized case of infiltration of natural rain with a time-variable intensity into the soil with depth-variable properties. This problem is the subject of the present paper.

The generalized model describes the infiltration process in two phases. During an unsaturated period without overland flow the entire rain water infiltrates and produces a new distribution of soil moisture. In the case of high rain intensities the soil surface is saturated, followed by a recession phase during which the infiltration intensity lessens and there occurs a surface runoff and overland flow as well as interflow at the layer boundaries.

We assume the formation of a distinct wetting front in the soil during a precipitation event, which separates the unsaturated soil with the antecedent moisture Θ_0 from the saturated part with the volumetric water content Θ_s. The refillable porosity $n = \Theta_s - \Theta_0$ can assume different values for the separate soil layers. This piston flow approach constitutes the decisive premise of the model, thus resulting in a simplified algorithm. The realistic moisture profile is approximated by a saturated column which is going deeper and deeper into the soil. If we exclude a growing volume of the soil, there is no storage possible in the tension-free saturated soil, which means that we can apply the DARCY equation to the saturated part of the step profile and neglect the continuity equation or only have to take it into account in a very simplified manner. This leads, in terms of the requirements mentioned above, to a reduced and physically founded model.

2. Calculation of the saturation phase

We assume a natural or artificial rain that is described by a hydrograph with a discrete time-variable rain intensity (cf. Fig. 1):

$$\mathrm{PI}(t) = \mathrm{PI}_v = \text{const} \quad \text{for } (v - 1)\,\Delta t \leqq t < v \cdot \Delta t \quad \text{and } v = 1, 2, \ldots, \mathrm{n} \tag{1}$$

PI(t) is the time-variable rain intensity and Δt the duration of the chosen time interval. Natural soil profiles in catchment areas are seldom homogeneous. In general, the decisive physical parameters of the soil are functions of the depth. Then the real profile is conveniently approximated by a number of layers with the parameters within each layer being constant (Fig. 1).

In dependence on the relation between the rain intensity PI_v and the saturated hydraulic conductivities K_i of the individual layers, the wetting front may generally

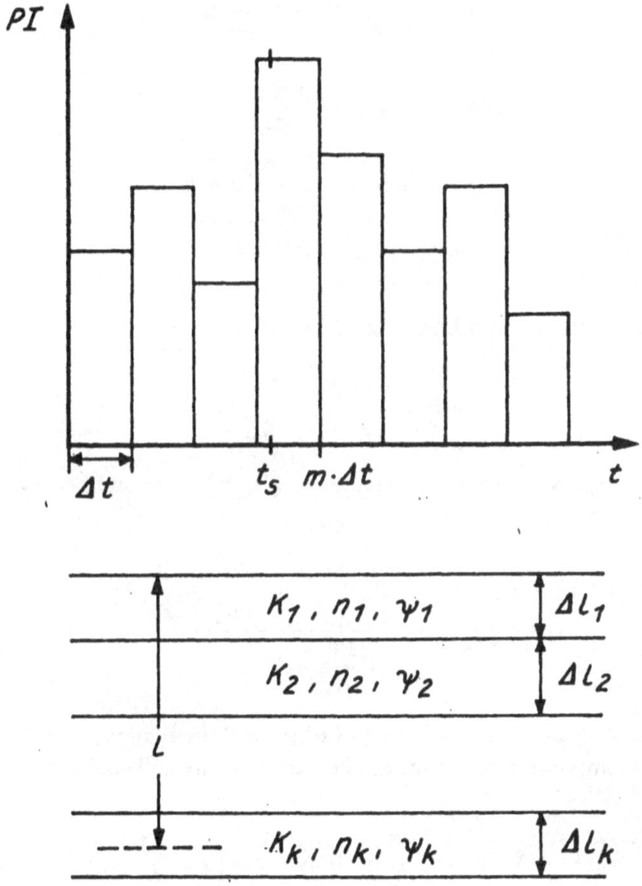

Fig. 1. Discrete rain hyetograph and layered soil profile. At the time of ponding the wetting front is within the kth layer

have penetrated into the kth layer until the beginning of the overland flow. Δl_i is to be the individual thickness of the soil layers, ψ_i the corresponding suction heads and PI_m the rain intensity of the mth time interval that leads to the saturation of the soil surface. We assume that at that moment the wetting front is in the kth layer. We proceed from the premise that the water meets the kth layer with a filter velocity that is equal to the value of the rain intensity PI_m even if the surface saturation is not reached during the passage of the wetting front through the k-1 layers above it. If we apply to the saturated sublayers of the soil profile the DARCY law, we will get the following equation system:

$$\mathrm{PI}_m = K_1 \frac{\Delta l_1 + \psi_1 - \psi_0}{\Delta l_1}$$

$$\mathrm{PI}_m = K_2 \frac{\Delta l_2 + \psi_2 - \psi_1}{\Delta l_2}$$

$$\mathrm{PI}_m = K_k \frac{l - \sum_{i=1}^{k-1} \Delta l_i + \psi_k - \psi_{k-1}}{l - \sum_{i=1}^{k-1} \Delta l_i}. \tag{2}$$

We solve equation (2) according to ψ_{k-1} by successive elemination of ψ_1, ψ_2, etc.:

$$\psi_{k-1} = \mathrm{PI}_m \cdot \sum_{i=1}^{k-1} \frac{\Delta l_i}{K_i} - \sum_{i=1}^{k-1} \Delta l_i + \psi_0 . \tag{3}$$

If we use this expression for ψ_{k-1} in equation (2), the depth attained by the wetting front will be

$$l = \frac{\psi_k - \psi_0 + \mathrm{PI}_m \left[\sum_{i=1}^{k-1} \Delta l_i \left(\frac{1}{K_k} - \frac{1}{K_i} \right) \right]}{\frac{\mathrm{PI}_m}{K_k} - 1}. \tag{4}$$

The infiltration quantity F until the end of the mth time interval can be expressed both by the temporal summation of the rain and the soil moisture deficit filled up to depth l:

$$F = \sum_{\nu=1}^{m} \Delta t_\nu \, \mathrm{PI}_\nu = \sum_{i=1}^{k-1} n_i \, \Delta l_i + n_k \left(l - \sum_{i=1}^{k-1} \Delta l_i \right), \tag{5}$$

l is therfore

$$l = \frac{1}{n_k} \left[\sum_{\nu=1}^{m} \Delta t_\nu \, \mathrm{PI}_\nu + \sum_{i=1}^{k-1} \Delta l_i (n_k - n_i) \right]. \tag{6}$$

Equating (4) and (6) provides:

$$\psi_0 = \psi_k + \mathrm{PI}_m \left[\sum_{i=1}^{k-1} \Delta l_i \left(\frac{1}{K_k} - \frac{1}{K_i} \right) \right] - \frac{1}{n_k} \left(\frac{\mathrm{PI}_m}{K_k} - 1 \right) \left(\sum_{\nu=1}^{m} \Delta t_\nu \, \mathrm{PI}_\nu + \sum_{i=1}^{k-1} \Delta l_i (n_k - n_i) \right). \tag{7}$$

At the moment when the soil surface is saturated ψ_o disapperars. If we insert in (7) $\psi_0 = 0$, the saturation will be reached at the end of the mth interval, because in (7) the temporal summation continues until m is reached. If saturation is reached within the mth interval, $\psi_o < 0$ is applicable and (7) leads to

$$\left(\frac{PI_m}{K_k} - 1\right)\left(\sum_{v=1}^{m} \Delta t_v PI_v + \sum_{i=1}^{k-1} \Delta l_i (n_k - n_i)\right) > n_k \left[\psi_k + PI_m \sum_{i=1}^{k-1} \Delta l_i \left(\frac{1}{K_k} - \frac{1}{K_i}\right)\right]. \tag{8}$$

With (8) a condition has been derived that permits us to determine the time interval m in which the soil surface is saturated.
For the homogeneous soil $n_k = n_i = n$, $K_k = K_i = K$ and $\psi_k = \psi$ is applicable, and (8) is reduced to

$$\left(\frac{PI_m}{K} - 1\right)\left(\sum_{v=1}^{m} \Delta t_v PI_v\right) > \psi \cdot n . \tag{8 a}$$

In order to fix the concrete ponding time t_s within the mth interval, we basically continue the previous derivation, but we have to take into consideration that in (5) the summation may only go up to $m - 1$. The resultant expression for l runs

$$l = \frac{1}{n_k}\left(\sum_{v=1}^{m-1} \Delta t_v PI_v + PI_m\left(t - \sum_{v=1}^{m-1} \Delta t_v\right) + \sum_{i=1}^{k-1} \Delta l_i (n_k - n_i)\right). \tag{9}$$

If we now equate (9) with (4) and consider that at the time when the soil surface is saturated $t = t_s$ and $\psi_0 = 0$, the ponding time will be

$$t_s = \sum_{v=1}^{m-1} \Delta t_v \left(1 - \frac{PI_v}{PI_m}\right) + \frac{1}{PI_m} \sum_{i=1}^{k-1} \Delta l_i (n_i - n_k) + \frac{n_k \left[\psi_k + PI_m \sum_{i=1}^{k-1} \Delta l_i \left(\frac{1}{K_k} - \frac{1}{K_i}\right)\right]}{PI_m \left(\frac{PI_m}{K_k} - 1\right)}. \tag{10}$$

For equally big time intervals Δt is reduced (10) to

$$t_s = (m-1)\,\Delta t - \frac{\Delta t}{PI_m} \sum_{v=1}^{m-1} PI_v + \frac{1}{PI_m} \sum_{i=1}^{k-1} \Delta l_i (n_i - n_k) + \frac{n_k \left[\psi_k + PI_m \sum_{i=1}^{k-1} \Delta l_i \left(\frac{1}{K_k} - \frac{1}{K_i}\right)\right]}{PI_m \left(\frac{PI_m}{K_k} - 1\right)}. \tag{10 a}$$

For the homogeneous soil it follows

$$t_s = (m - 1)\,\Delta t - \frac{\Delta t}{\mathrm{PI}_m} \sum_{v=1}^{m-1} \mathrm{PI}_v + \frac{n\psi}{\mathrm{PI}_m \left(\frac{\mathrm{PI}_m}{K} - 1\right)} \tag{10b}$$

and, finally, for constant rain intensity

$$t_s = \frac{Kn\psi}{\mathrm{PI}_m(\mathrm{PI}_m - 1)} . \tag{10c}$$

Summing up, we can conclude with regard to (8) and (10) that the duration of the saturation phase and along with it the question whether and when there is runoff formation, essentially depends on the correlation between the rain intensity and the conductivity of the soil, on the wetting during rainfall and on the moisture deficit in the soil at the beginning of the rainfall.

Due to the character of the discrete rain hydrograph may be

$$t_s = (m - 1)\,\Delta t \tag{11}$$

i.e. the saturation of the soil is achieved at the time of the rapid increase in rainfall intensity at the beginning of the mth time interval. If we want to derive the condition for making a decision between (10) and (11), all terms of the right side except the first one are set to zero. In this new equation PI_m has just the value to place saturation at the beginning of the mth interval. PI_{gr} designates a limit value of the precipitation intensity, to which the following equation is applicable:

$$\mathrm{PI}_{gr} = K_k \frac{a + n_k \psi_k}{a - n_k K_k \sum_{i=1}^{k-1} \Delta l_i \left(\frac{1}{K_k} - \frac{1}{K_i}\right)} \tag{12}$$

with

$$a = \Delta t \sum_{v=1}^{m-1} \mathrm{PI}_v + \sum_{i=1}^{k-1} \Delta l_i (n_k - n_i) . \tag{13}$$

In other words: If

$$\mathrm{PI}_m \geqq \mathrm{PI}_{gr} , \tag{14}$$

the ponding time is calculated according to (11), for $\mathrm{PI}_m < \mathrm{PI}_{gr}$ or $\mathrm{PI}_{gr} < 0$ the equation (10a) has to be used.

The depth reached by the wetting front at the ponding time, the so-called saturation depth l_s is calculated with $\psi_0 = 0$ from (4):

$$l_s = \frac{\psi_k + \mathrm{PI}_m \sum_{i=1}^{k-1} \Delta l_i \left(\frac{1}{K_k} - \frac{1}{K_i}\right)}{\frac{\mathrm{PI}_m}{K_k} - 1} . \tag{15}$$

The infiltration quantity up to saturation can be expressed in analogy to (5) in two ways:

$$F_s = n_k l_s + \sum_{i=1}^{k-1} \Delta l_i (n_i - n_k) \tag{16}$$

or

$$F_s = \mathrm{PI}_m [t_s - (m-1)\,\Delta t] + \Delta t \sum_{\nu=1}^{m-1} \mathrm{PI}_\nu . \tag{16 a}$$

If in (16a) or (16b) the values for l_s are inserted from equation (15) and for t_s from equation (10a), the result will be

$$F_s = \frac{n_k \left[\psi_k + \mathrm{PI}_m \sum_{i=1}^{k-1} \Delta l_i \left(\frac{1}{K_k} - \frac{1}{K_i}\right)\right]}{\frac{\mathrm{PI}_m}{K_k} - 1} + \sum_{i=1}^{k-1} \Delta l_i (n_i - n_k) \tag{16b}$$

A comparison between (16b) and (8) provides for constant Δt

$$\Delta t \sum_{\nu=1}^{m} \mathrm{PI}_\nu > F_s \tag{17}$$

So we have attained a very clear form of the saturation condition. It says that the rainfall during m intervals must be greater than the quantity infiltrated during this time, if saturation is to occur in the mth interval.

The test (14) can be illustrated in a similar way. From (14) we get with the aid of (12) and after minor conversions

$$\Delta t \sum_{\nu=1}^{m-1} \mathrm{PI}_\nu = \frac{n_k \left[\psi_k + \mathrm{PI}_m \sum_{i=1}^{k-1} \Delta l_i \left(\frac{1}{K_k} - \frac{1}{K_i}\right)\right]}{\frac{\mathrm{PI}_m}{K_k} - 1} + \sum_{i=1}^{k-1} \Delta l_i (n_i - n_k) \tag{18}$$

The comparison between (16b) and (18) leads to

$$\Delta t \sum_{\nu=1}^{m-1} \mathrm{PI}_\nu = F_s \,. \tag{19}$$

For $\mathrm{PI}_{gr} > \mathrm{PI}_m$ the result will be

$$\Delta t \sum_{\nu=1}^{m-1} \mathrm{PI}_\nu < F_s \,. \tag{19a}$$

We thus have evidence the cumulative infiltration until ponding time calculated according to (16b) must be greater at the beginning of or within the mth interval than or equal to the rainfall until the beginning of this interval.

We are now able to determine all important magnitudes of the saturation phase for the case that sufficiently large rainfall intensities and/or a sufficiently long lasting rain lead to the saturation of the soil profile under consideration.

3. Increase in soil moisture

The saturation phase described so far is sufficiently dealt with if the calculation of the infiltration enables us to determine the effective precipitation of a single high-water event. Our present objective moreover requires providing information about the increase in soil moisture in the separate layers following rain events which do not lead to the saturation of the soil surface and thus do not lead to overland flow.

With $\psi_0 = 0$ in (7) PI_m has just the value that results in the saturation of the soil surface at the end of the mth interval. We call this rain intensity $\mathrm{PI}_{\text{pot},\nu}$, because we can calculate it for each time interval. If we put in (7) $\psi_0 = 0$, $\mathrm{PI}_m = \mathrm{PI}_{\text{pot},\nu}$, $\Delta t_\nu = \Delta t = \text{const}$ and use the abbreviations

$$b = \sum_{i=1}^{k-1} \Delta l_i (n_k - n_i)$$

and

$$c = \sum_{i=1}^{k-1} \Delta l_i \left(\frac{1}{K_k} - \frac{1}{K_i}\right),$$

the resultant equation will be

$$\frac{1}{n_k}\left(\frac{\mathrm{PI}_{\mathrm{pot},\nu}}{K_k}-1\right)\left(\Delta t\sum_{\nu=1}^{m-1}\mathrm{PI}_\nu+\Delta t\cdot\mathrm{PI}_{\mathrm{pot},\nu}+b\right)=\psi_k+\mathrm{PI}_{\mathrm{pot},\nu}\cdot c\,. \tag{20}$$

For $\mathrm{PI}_{\mathrm{pot},\nu}$ we derive from (20) the determining equation

$$\mathrm{PI}_{\mathrm{pot},\nu}^{\;2}+\mathrm{PI}_{\mathrm{pot},\nu}\cdot d-e=0 \tag{21}$$

with

$$d=\sum_{\nu=1}^{m-1}\mathrm{PI}_\nu+\frac{b}{\Delta t}-K_k-\frac{K_k n_k c}{\Delta t}$$

and

$$e=\frac{K_k}{\Delta t}\left(\Delta t\sum_{\nu=1}^{m-1}\mathrm{PI}_\nu+b+n_k\psi_k\right).$$

The physically correct solution of (21) is

$$\mathrm{PI}_{\mathrm{pot},\nu}=-\frac{d}{2}+\sqrt{\frac{d^2}{4}+e}\,. \tag{22}$$

The potential rainfall intensity $\mathrm{PI}_{\mathrm{pot},\nu}$ causes a saturated soil profile with the thickness $l_{\mathrm{pot},\nu}$ in accordance with the assumptions of the model. In analogy to (15) we obtain

$$l_{\mathrm{pot},\nu}=\frac{\psi_k+\mathrm{PI}_{\mathrm{pot},\nu}\sum_{i=1}^{k-1}\Delta l_i\left(\frac{1}{K_k}-\frac{1}{K_i}\right)}{\frac{\mathrm{PI}_{\mathrm{pot},\nu}}{K_k}-1} \tag{23}$$

All rainfall intensities $\mathrm{PI}_\nu>\mathrm{PI}_{\mathrm{pot},\nu}$ lead to the saturation of the soil surface during the time interval $\nu=m$. Rainfall intensities that are smaller than $\mathrm{PI}_{\mathrm{pot},\nu}$ do not saturate the soil surface, they only increase the soil moisture until depth $l_{\mathrm{pot},\nu}$ to the value $\Theta_a<\Theta_s$. We obtain Θ_a from

$$\Theta_a=\Theta_s-\frac{100\cdot\Delta t\,(\mathrm{PI}_{\mathrm{pot},\nu}-\mathrm{PI}_\nu)}{l_{\mathrm{pot},\nu}}\,. \tag{24}$$

4. Calculation of the recession phase

As long as the soil surface is not yet saturated by water, the entire rain water infiltrates, which means that the infiltration intensity is equal to the rainfall intensities PI_v during the saturation phase. This no longer applies to the times $t > t_s$ during the recession phase [3]. If we assume the wetting front during the the recession phase is in the nth layer, in equation (4) $\psi_5 = 0$ and $PI_m = f_n$, with f_n describing the infiltration intensity with which the water moves in the nth layer. With these modifications is follows through minor transformations of (4):

$$f_n = \frac{l + \psi_n}{\frac{l}{K_n} + \sum_{i=1}^{n-1} \Delta l_i \left(\frac{1}{K_i} - \frac{1}{K_n}\right)} . \tag{25}$$

The velocity of the wetting front is calculated – taking into account the general relationship between true velocity and filter velocity – from

$$\frac{dl}{dt} = \frac{f_n}{n_n} . \tag{26}$$

From (25) and (26) we get this differential equation

$$dt = n_n \frac{\frac{l}{K_n} + \sum_{i=1}^{n-1} \Delta l_i \left(\frac{1}{K_i} - \frac{1}{K_n}\right)}{l + \psi_n} dl . \tag{27}$$

Its integration in the layer limits for the time needed by the wetting front to penetrate the nth layer leads to

$$\Delta t_n = \frac{n_n}{K_n} \left\{ \Delta l_n + \ln \frac{\psi_n + \sum_{i=1}^{n} \Delta l_i}{\psi_n \sum_{i=1}^{n-1} \Delta l_i} \left(\sum_{i=1}^{n-1} \Delta l_i \frac{K_n}{K_i} - 1 \right) - \psi_n \right\} . \tag{28}$$

Considering the relationship between length and infiltration quantity for the nth layer

$$\Delta F_n = n_n \cdot \Delta l_n \tag{29}$$

the relationship between time and infiltration quantity can also be simply derived from (28):

$$\frac{K_n}{n_n}\,\Delta t_n = \frac{\Delta F_n}{n_n} + \ln \frac{\psi_n + \sum\limits_{i=1}^{n} \frac{\Delta F_i}{n_i}}{\psi_n + \sum\limits_{i=1}^{n-1} \frac{\Delta F_i}{n_i}} \left[\sum_{i=1}^{n-1} \frac{\Delta F_i}{n_i}\left(\frac{K_n}{K_i} - 1\right) - \psi_n\right]. \tag{28a}$$

The relationship between infiltration F and time t is unfortunately given only implicitly by (28) and (29). One possibility to arrive at the explicit dependence $F(t)$ is to approximate the logarithm, e.g. by

$$\ln x \approx 2\,\frac{x-1}{x+1}, \tag{30}$$

with (30) representing only the first term of the series expansion of the natural logarithm. This approximation is very rough and can be justified only for practical calculations: $x < 2$ in general, that means that the errors of the approximation remain inferior to 4 per cent. With the aid of (30) we derive from (28 a) the following determining equation:

$$F_n^2 + F_n\left(2\sum_{i=1}^{n-1} \frac{\Delta F_i}{n_i} \cdot \frac{K_n n_n}{K_i}\right) - F_n K_n t_n - 2K_n n_n t_n \left(\psi_n + \sum_{i=1}^{n-1} \frac{\Delta F_i}{n_i}\right) = 0\,. \tag{31}$$

In (31) ΔF_n is substituted by F_n and Δt_n by t_n as against (28 a). This is to emphasize that any interim values of the function $F(t)$ can be calculated as long as the wetting front moves in the nth layer. The solution of (31) is

$$F_n = At_n - B + \sqrt{(At_n - B)^2 + 2ACt_n} \tag{32}$$

with

$$A = \frac{K_n}{2} \tag{33}$$

$$B = n_n K_n \sum_{i=1}^{n-1} \frac{\Delta l_i}{K_i} \tag{34}$$

$$C = 2n_n\left(\psi_n + \sum_{i=1}^{n-1} \Delta l_i\right). \tag{35}$$

This explicit dependence $F(t)$ offers the chance to derive

$$f(t) = \frac{\mathrm{d}F(t)}{\mathrm{d}t}$$

as an analytical expression. The result of the differentiation is

$$f_n(t) = A\left\{\frac{(At_n - B) + C}{\sqrt{(At_n - B)^2 + 2ACt_n}} + 1\right\}. \tag{36}$$

The equations (32) and (36) only indicate the real infiltration course for the time in which the wetting front is within the nth layer. In the case of layer-variable parameters (cf. equations (33) to (35)), a different infiltration curve $f_i(t)$ can be calculated according to (36) for each layer. The entire course of infiltration in the recession phase is composed of these $f_i(t)$. If we want to formulate the conditions following from this, we must know the times when the wetting front arrives at and leaves again the layer under consideration. Proceeding from (28) and with the aid of (30) this transition time t_n can be derived for the nth layer:

$$t_n = \frac{D}{K_n}\left(1 + \frac{2B - C}{D + C}\right), \tag{37}$$

where

$$D = n_n \cdot l_n \tag{38}$$

and B and C from equation (34) and (35) must be inserted.

The first resultant condition is given by the fact that at ponding time $t = t_s$ the cumultative infiltration must be $F = F_s$. Taking into consideration (32), we get for cumulative infiltration

$$\mathrm{F}(t) = F_s + A(t - t_s) - B + \sqrt{(A(t - t_s) - B)^2 + 2AC(t - t_s)} \tag{32a}$$

and for the temporal dependence of the infiltration intensity

$$f(t) = A\left\{1 + \frac{A(t - t_s) - B + C}{\sqrt{(A(t - t_s) - B)^2 + 2AC(t - t_s)}}\right\}. \tag{36a}$$

If we take into account that at the entry of the wetting front into a new layer the transition time t_n must be added to t_s in (32a) und (36a), and the parameters A to D must be newly calculated, it is possible to calculate with (32a) and (36a) the functions $F(t)$ and $f(t)$ for the entire recession phase.

If, with the aid of (36a) and certain layer parameters $f(t)$ is calculated formally outside the considered layer (in contrast to the real infiltration curve we call these parts of $f(t)$ fictitious), then the definition area of $f(t)$ to be derived from (36a) is also of interest:

$$t_s + \frac{1}{A}\left[B - C\left(1 - \sqrt{1 - \frac{2B}{C}}\right)\right] < t < \infty . \tag{39}$$

5. Determination of the runoff components

Supposing that the rainfall intensity is always greater than or equal to the infiltration intensity, the recession phase of infiltration is completely described by (32a) or (36a). In this case the water quantity shown above the infiltration curve $f(t)$ forms an overland flow RO, the amount of which is calculated as follows:

$$\mathrm{RO}_\nu = \mathrm{P}_\nu - \mathrm{F}_\nu , \tag{40}$$

with F_ν being derived from (32a) for $(\nu - 1)\,\Delta t \leqq t < \nu\,\Delta t$.
Frequently, however, the case $\mathrm{PI}(t) < f(t)$ occurs during natural precipitation events. The soil could absorb more water than it is offered. A saturation deficit will result:

$$D_\nu = \mathrm{F}_\nu - \mathrm{P}_\nu . \tag{41}$$

Another precipitation surplus and hence an overland flow will only be possible when the deficit is balanced following another increase in rain intensity. The procedure to be applied in this case constitutes a continuous balancing of precipitation and precipitation surplus, related to the $f(t)$ curve determined before.

If the infiltration front enters a new soil layer with lower values of the physical parameters, infiltration conditions will deteriorate, and $v_{i+1} < v_i$ will apply for the speed at which the wetting front moves. At the lower limit of the ith layer there occurs water accumulation or a potential interflow amount which runs off with a time delay at an adequate slope of stratum. In order to ascertain the interflow RH, we need fictitious and real characteristics of infiltration of successive layers, since RH will be calculated as the difference between the fictitious value $\mathrm{F}_{\nu,i}$ and the real value $\mathrm{F}_{\nu,i+1}$:

$$\mathrm{RH}_\nu = \mathrm{F}_{\nu,i} - \mathrm{F}_{\nu,i+1} . \tag{42}$$

6. One example

In conclusion we should like to illustrate the calculation by presenting an example. We will consider a storm event whose discrete hydrograph is shown in Fig. 2.

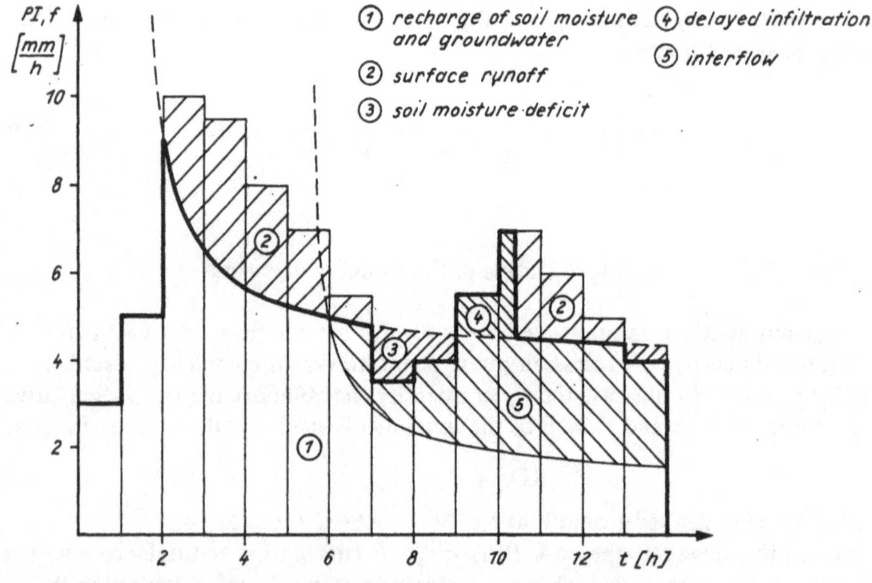

Fig. 2. Presentation of a storm hyetograph and the division of precipitation water into runoff components, soil moisture and groundwater recharge

It meets a small catchment area whose soil mainly consists of sandy to argillaceous loam and can be represented as a two-layer profile. The ascertained parameters are $\Delta l_1 = 320$ mm, $K_1 = 4 \text{ mm} \cdot \text{h}^{-1}$, $\psi_1 = 100$ mm, $n_1 = 0.1$; $\Delta l_2 = 320$ mm, $K_2 = 1 \text{ mm} \cdot \text{h}^{-1}$, $\psi_2 = 50$ mm, $n_2 = 0.05$.

1 We determine the precipitation interval m, in which the soil surface is saturated for the first time, by gradually examining test (8), and asertain $m = 3$ and $k = 1$, i.e. the wetting front is still in the first layer at the moment of saturation. This solution also results from the calculation of the potential precipitation intensity according to (22), since for the first time $\text{PI}_{\text{pot},3} < \text{PI}_3$ ($\text{PI}_{\text{pot},3} = 6.7 \text{ mm} \cdot \text{h}^{-1}$, $\text{PI}_3 = 10 \text{ mm} \cdot \text{h}^{-1}$) is applicable to the third time interval.

2 In accordance with (12) $\text{PI}_{\text{gr}} = 9 \text{ mm} \cdot \text{h}^{-1}$ is calculated. It is therefore $\text{PI}_m > \text{PI}_{\text{gr}}$, which means that with (11) $t_s = 2$ h.

3 According to (16) and (16a) we calculate $\text{F}_s = 8$ mm and $l_s = 80$ mm. Thus the conditions (17) and (19) have been fulfilled.

4 Making use of (22) to (24) we obtain an increase in the antecedent moisture from 40 to 43.5 per cent per volume after the first time interval and to 47.5 per cent per volume after the second time interval.

5 The time the wetting front needs to get from the saturation depth of $l_s = 80$ mm to the end of the first layer at $\Delta l_1 = 320$ mm, is calculated according to equation (37) and is $t_1 = 4$ h.
6 In accordance with (32a) and (36a) we calculate the cumulative infiltration F(t) and the infiltration curve $f(t)$, which is depicted in Fig. 2.
7 According to equation (40) an overland flow occurs in the time intervals 3 to 7 and 11 to 13. Its total sum is 14.8 mm. In the intervals 8 and 9 there is a soil moisture deficit of 1.9 mm, which is not completely balanced until the rainfall in the 11th time interval. The sum of the interflow which is formed in the intervals 7 to 14 is 17 mm.
8 The entire precipitation thus consits of the overland flow (17.8 per cent), the interflow (20.5 per cent) and an increase in soil moisture and ground water recharge (61.7 per cent).

References

[1] Bouwer, H.: Infiltration of Water into Nonuniform Soil. — J. Irrig. Drain. Div. **95** (1969) IR4, 451—462.

[2] Childs, E. C.: An Introduction to the Physical Basis of Soil Water Phenomena. — London/New York: J. Wiley & Sons, 1969.

[3] Dyck, S.; Peschke, G.: Grundlagen der Hydrologie. — Berlin: Verlag für Bauwesen, 1983.

[4] Mein, R. G.; Larson, C. L.: Modeling Infiltration during a Steady Rain. — Water Resour. Res. **9** (1973) 2, 384—394.

[5] Neuman, S. P.: Wetting Front Pressure Head in the Infiltration Model of Green and Ampt. — Water Resour. Res. **12** (1976) 3, 564—566.

[6] Onstad, C. A.; Olsen, T. C.; Stone, L. R.: An Infiltration Model Tested with Monolith Moisture Measurements. — Soil Sci. **116** (1973) 1, 13—17.

[7] Peschke, G.: Möglichkeiten und Grenzen bei der mathematischen Modellierung der Infiltration. — Geod. Geoph. Veröff. R. IV, H. **20** (1976), 74—83.

[8] Peschke, G.: Ein zweistufiges Modell der Infiltration von Regen in geschichtete Böden. — Acta hydrophys. **22** (1977) 1, 39—48.

[9] Peschke, G.: Strukturanalyse des Zweistufenmodells zur Berechnung der Infiltration. — Wiss. Z. Techn. Univers. Dresden **28** (1979) 6, 1603—1607.

[10] Peschke, G.: Die Infiltration während eines zeitvariablen Regens. — Geod. Geoph. Veröff. R. IV, H. **32** (1980), 120—125.

[11] Peschke, G.: Zur Bildung und Berechnung von Regenabfluß. — Wiss. Z. Techn. Univers. Dresden **34** (1985) 4, 195—200.

[12] Peschke, G.; Gurtz, J.: The role of Infiltration in Rainfall-Runoff-Relationship. — Acta hydrophys. **31** (1987) 3/4, 207—223.

[13] Swartzendruber, D.: Infiltration of Constant-Flux Rainfall into Soil as Analyzed by the Approach of Green and Ampt. — Soil. Sci. **117** (1974) 5, 272—281.

Acta hydrophys., Berlin **31** (1987) 3/4, S. 207–223

The Role of Infiltration in Rainfall-Runoff Relationship

By G. PESCHKE and J. GURTZ[1])

Summary The infiltration is not only a key variable for all the problems connected with the determination of soil moisture, interflow and groundwater recharge, but it is also very important to the hydrologic analysis of precipitation-runoff relationhips. These different problems of application necessitate also solutions of the infiltration problem with different degrees of complexity Alternative solutions are necessary, however, also since none of them is capable of taking into account all the quantities influencing infiltration.

The subject of this paper consists in discussing the role of infiltration concerning the quantitative description of precipitation-runoff relationships. This task has been treated in hydrology frequently with the aid of empirical infiltration equations in different ways and without a soil-physics-based algorithm. From our experimental investigations carried out for many years on infiltration for the mentioned hydrological application, the intensity of precipitation and the antecedent soil moisture proved to be the main quantities influencing the infiltration process. Proceeding from the basic assumption that the infiltration rate is a function only of the increasing storage of soil moisture, an algorithm is derived within which a soil-physics-based generalization of the known empirical infiltration equations is made and their uniform application is possible, and which includes the mentioned main influencing quantities. This algorithm is the basis of a simple infiltration model for describing precipitation-runoff relationships.

Zusammenfassung Die Infiltration nimmt nicht nur eine Schlüsselstellung für alle Probleme im Zusammenhang mit der Bestimmung von Bodenfeuchtigkeit, hypodermischem Abfluß und Grundwasserneubildung ein, sie ist ebenso von großer Bedeutung für die hydrologische Analyse der Niederschlag-Abfluß-Beziehungen. Derart unterschiedliche Anwendungsgebiete erfordern unterschiedlich komplexe Lösungen des Infiltrationsproblems, was, da keine dieser Lösungen alle die Infiltration beeinflussenden Größen zu erfassen vermag, alternative Lösungen nötig macht. Gegenstand der vorliegenden Arbeit ist die Diskussion der Rolle der Infiltration für eine quantitative Beschreibung von Niederschlag-Abfluß-Prozessen Diese Prozesse wurden in der Hydrologie bereits früher verschiedentlich mit Hilfe empirischer Infiltrationsgleichungen auf

[1]) Prof Dr sc. nat. GERD PESCHKE, Dr rer nat. JOACHIM GURTZ, Technische Universität Dresden, Sektion Wasserwesen, Bereich Hydrologie und Meteorologie, Mommsenstraße 13, Dresden, DDR-8027

unterschiedlichen Wegen, aber außerhalb eines bodenphysikalisch begründeten Algorithmus behandelt. Unsere langjährigen experimentellen Untersuchungen der Infiltrationsproblematik ergaben, daß die Intensität der Niederschläge und die Bodenfeuchte vor dem Niederschlagsereignis als die Haupteinflußgrößen für den Infiltrationsprozeß anzusehen sind. Ausgehend von der Grundanahme, daß die Infiltrationsintensität nur eine Funktion der zunehmenden Feuchte im Bodenprofil ist, wurde ein bodenphysikalisch begründeter Algorithmus abgeleitet, der durch Verallgemeinerung der bekannten empirischen Infiltrationsgleichungen deren einheitliche Anwendung ermöglicht und, der die erwähnten Haupteinflußgrößen enthält. Dieser Algorithmus ist die Grundlage für ein einfaches Infiltrationsmodell zur Beschreibung von Niederschlag-Abfluß-Beziehungen.

1. Introduction

By infiltration we understand the entry of water into the soil through its surface, and because of the hydrologic importance and for reasons of the confinement on a partial topic we will refer to the infiltration of rain only. The infiltration process is of such a great importance because it splits the supply from precipitation into surface and subsurface water at a central position of the hydrologic cycle. The fact is reflected in the prominent position of the infiltration model within complex watershed models. It is mainly due to the condition that the majority of the processes to be modelled in the soil, on the land surface and in the river bed occur after infiltration, so that the quality of the results depends very much on the accuracy of the input quantities supplied by the infiltration model for the submodels.

The infiltration is the decisive input process into the aeration zone system, being connected organically with it. The neglection of the feedback of processes in the unsaturated soil to infiltration was the main cause of the fact that in hydrology it was described for decades as a pure interfacial process, i.e. insufficiently and without any physical foundation.

As to its content, the infiltration process constitutes the connecting link between surface and subsurface water, with respect to methodology its place is between hydrology and soil physics, and regarding the applications of solutions it is of interest likewise for water resources management, agriculture and forestry, for civil engineering, power economy and the transport system. The demands made by different users on solutions of the infiltration problem, however, are different. On the one hand, for the analysis and prediction of short-term precipitation-runoff relationships only the division of rain into overland flow and subsurface water is of interest; on the other hand, there are of greater interest the dynamics of moisture in the soil profile and its influence on the mechanical, thermal, electrical, biological and chemical soil properties. For that reason alone, the solutions should have different degrees of complexity: besides solutions of the complete nonlinear differential equations, simple equations should be available.

2. Process complexity and main influencing quantities

Alternative solutions are necessary also for the reason that no solution alone can reflect the complexity of the process. The infiltration is controlled by the quantity of water being available to the soil and by the transport properties in the soil profile. The available quantity depends on the properties of the process of precipitation and of the soil surface inclusive of the vegetation cover, the transport properties being determined by soil-physical parameters. Not only because of the resulting diversity of influencing factors but also because of their interdependence it is difficult to isolate the effect of individual quantities unambiguously. The difficulties within this complex of questions are reflected by a great variety of contradictory opinions with respect to the relative weights and partly also to the directions of action of the different influencing factors. In many cases, these contrasting conceptions are due to different methods and conditions of investigation. This is to be illustrated by a few examples.

For instance, differences in the soil properties are completely concealed when we carry out infiltration tests on plots with biologically created macropores or if the sites investigated are characterized by a greater variance of vegetation or soil structure due to agriculture. The infiltration-inhibiting influence or urbanization cannot be significantly proved for heavy soils with a continuously high antecedent soil moisture, since due to that, anyhow, the investigation is restricted to the range of the lowest infiltration capacity. With double ring infiltrometers influence neither of the rain intensity nor of the slope of the soil surface can be determined. The infiltration in closed columns will be influenced always by the countercurrent of air, whereas this occurs only rarely during measurements in the field.

Contradictory statements are caused also by lacking precision. For example, a statement as: "Growing rain intensity increases (or reduces) infiltration" does not make any physical sense. As long we interpret the rain intensity in the sense of boundary conditions of the second type, it will be effective as an influencing factor only during the unsaturated period of the process of infiltration. Here, it has an increasing effect on the infiltration rate and a reducing one on the cumulative infiltration, since its influence on the reduction of the ponding time is of a higher order as compared with the increase of the infiltration rate. If, on the other hand, more attention is paid to the energy transferred to the surface of bare soils by the size and rate of fall of rain-drops, the rain intensity is effective also during the recession phase of infiltration due to the destruction of aggregated soil particles with subsequent sludging and compaction; in this case, however, the infiltration rate reduced by it. In a similar way, it is physically not correct to explain the influence of the rain intensity on infiltration by the change of parameters, e.g. of the HORTON equation [12], if the equation applies only under boundary conditions of the first type.

The difficult-to-disentangle complex in which the diversity of the influencing factors may combine under certain conditions indicates the limits which are set to the determinist modelling of the process of infiltration. It is an unrealistic aim if one intends to take into account the greatest possible number of influencing quantities. Apparently, only the questions regarding the quantities exercising the main effect on the process and about the respective conditions are essential.

In this paper, we confine ourselves only on the role of infiltration in the dividing of rain-water into precipitation excess and subsurface water (PESCHKE, KUTILEK [25], paying no attention to the further distribution of the latter.
For other tasks the integration of the distribition of soil moisture demands a different approach to infiltration (PESCHKE et al. [26]). Such a restriction is typical for the analysis of shortterm precipitation-runoff relationships in hydrologic basins and was treated frequently by the use of empirical infiltration equations. By experimental investigations concerning this task, which were conducted for many years (PESCHKE, MÜLLER [24]), we found that among the larger number of factors influencing infiltration, the rain intensity and the refillable porosity of the soil or the antecedent soil moisture are of decisive importance. Therefore, both influencing quantities had to be taken into account in the algorithm for the calculation of infiltration for this task.

3. The model strategy

Each model is based on a theory on the system to be represented. An initial boundary value problem nonlinear partial second-order differential equations has to be solved for infiltration. Considering the expenditure required by such solutions, simplified models become necessary because of the legitimate desire for simplification and better adequacy to users. Such simplified or reduced models distinguish themselves by a lower content of physical information as against the complete solutions, but require a lower expenditure of simulation and are applicable to large systems. There should be emphasized, however, the use of complete solutions for the quantification of simplified models, since they permit their derivation in a science-based way, so that they will stand out from empirical trials, intuition or mathematical formalism. Our previous investigations (DYCK et al. [5]; PESCHKE [22]) showed that infiltration is described in sufficient agreement with the theoretical findings and the experimental experience if its description is based mainly on one of the initial physical equations only: on the DARCY equation or the continuity equation. This confirmed the conception held by CHILDS [1] that the infiltration rate can be described either as a consequence of the hydraulic conductivity and the gradient of the moisture potential in the soil surface or in dependence on the increase of the water stored in the soil.

4. Empirical infiltration equations

For the hydrologic task of calculating the precipitation excess it is sufficient that infiltration is known with respect to its temporal variation. This is accepted by many soil physicists, and therefore, they take part in the efforts made by hydrologists to derive simple infiltration equations and to use them hydrologically in the sense of the optimal compromise aimed at between the necessary simplicity and the possible soil-physical foundation. Since the number of these equations is growing, since they are very popular in hydrology due to their simplicity, and since the international difference of opinion concerning the most suitable of these equations continues, there is an urgent need of a critical valuation and of a recommendation for their uniform use.

The empirical infiltration equations can be classified into the group of power equations and the group of the exponential type. To the latter belong, e.g., the equations by HORTON [12] and POPOV [29], whereas part of the former are the equations by KOSTJAKOV [13], LEWIS [16], KUTERGIN [14], SOZYKIN [34], MEZENCEV [19], TISDALL [39], PHILIP [27], FOK [8], SMITH [32], SWARTZENDRUBER [37] and others, as will as the time-dependent representation of the equation by HOLTAN [10]. Special interest is due to the equations by PHILIP [27] as a simplified form of an analytical solution of the FOKKER-PLANCK equation and the equations by FOK [8], SMITH [32] and SWARTZENDRUBER [37] as adaptions of more complex solution functions. Thus, they may be regarded as early examples of the formation of reduced models.

The quotation of other authors would not result in any additional findings. It is the common aim of their equations to adapt measured results. From that also their common problems result:

- The empirical parameters have the task to ensure optimal adaptations. They can be determined from data of infiltration tests. Since they do not have an unambiguously definable connection with the physical process, they cannot be transferred to larger areas and are difficult to determine without infiltration experiments.
- These equations describe only the recession phase of infiltration or — in other words — they hold only for boundary conditions of the first type.
- Since their appearance, there have not been lacking attempts in the international literature to compare them with each other, to discuss their advantages and drawbacks, to decide for one of the equations for certain reasons, without even an approximate agreement having been achieved on a decision (e.g.,

FEDOROV [7]; PHILIP [28]; SWARTZENDRUBER, HUBERTY [36]; SKAGGS et al. [31]; LIN, GRAY [18]; SWARTZENDRUBER, YOUNGS [37], DUBREUIL [4]; CORRADINI, SINGH [2]. A true image of the controversy about the "best" of these equations is their application to complex watershed models: DAWDY, O'DONNELL [3] use the HORTON equation LICHTY et al. [17] apply PHILIP's equation, HOLTAN, LOPEZ [11] use the HOLTAN equation; SMITH, CHERRY [33] apply a modified MEZENCEV equation, whereas SIEGERT [30] and MOORE, LARSON [21] prefer the GREEN-AMPT approach (GREEN, AMPT [9]) — to mention only some typical examples. Submodels must be applied in almost all cases because of the inadequacy of the empirical infiltration equations.

In a summary survey (DYCK et al. [5], PESCHKE [22]) we discussed special and common features of empirical infiltration equations. Under the basic assumption that the temporal variation of the rate of infiltration can be satisfactorily calculated as a function of the water stored in the soil, the analysis permitted the following conclusions:

1. All the empirical infiltration equations can be derived and represented in a uniform way only under the assumption mentioned. By this, the variety of these equations has a uniform mathematical foundation.
2. Within error limits ($<4\%$) which are acceptable to hydrology. the equations give consistent results.
3. Their direct application to the calculation of the precipitation excess, however, is connected with large errors, since they describe only the recession phase of infiltration.
4. This drawback mentioned under 3. can be avoided by using the equations within an algorithm which permits their extension to the case of variable rain intensities. This algorithm is realized in the BLOINF infiltration model and it permits the uniform and hydrologically useful application of the empirical infiltration equations.

Thus, we are able to take a well-founded position on the diversity of opinions concerning the empirical infiltration equations: The controversy concerning the preferential choice of one or the other formula due to its good or less good adaptation to the measured variation of infiltration is unfruitful with respect to their hydrologic application. Because of the 2nd conclusion drawn above, we refer to equations which are as simple as possible.

5. The algorithm

Referring to previous investigations conducted in our institute (THEILE [38]) and to the confirmation by the work done by SMITH, CHERRY [33] and KUTILEK [15], we define the infiltration rate $f(t)$ by the following relation (PESCHKE [22]):

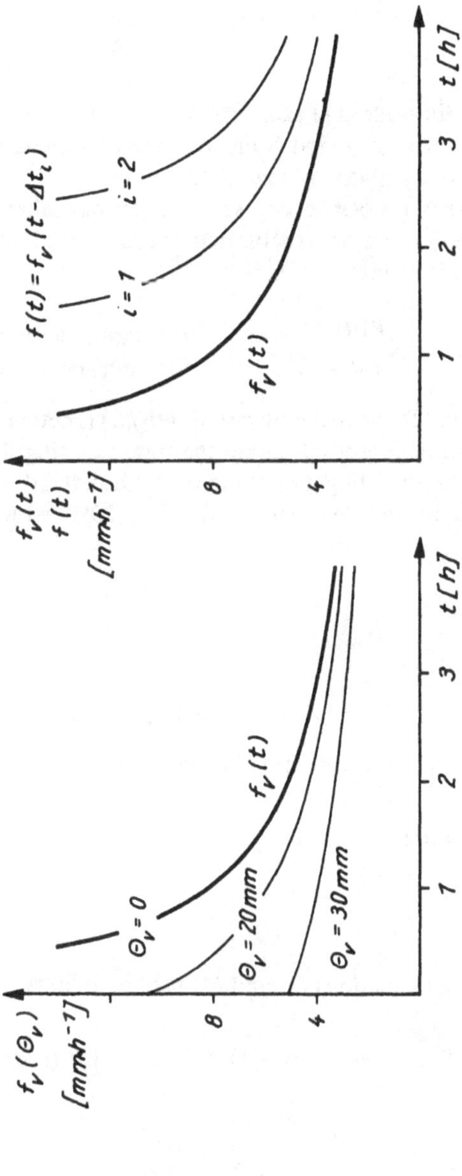

Fig. 1 The actual function of the infiltration rate $f(t)$ runs parallel to the infiltration capacity-time relationship of the dry soil ($f_v(t)$ or $f_v(\Theta_v = 0)$). The amount of shifting depends on antecedent soil moisture Θ_v (left) and on moistening during the rain (right)

$$f(t) = \begin{cases} \mathrm{PI}(t) & \text{for } t \leqq t_s \\ f_v(t - \Delta t') & \text{for } t > t_s \end{cases} \tag{1}$$

where $\mathrm{PI}(t)$ = the time-variable rain intensity, t = time, t_s = the ponding time separating the unsaturated period from the recession phase of infiltration, f_v = the infiltration capacity given by one of the empirical infiltration equations. If during a time-variable storm event serveral periods alternately succeed one another in which the soil surface is saturated or unsaturated, then the more general relationship (2) holds instead of (1):

$$f_i(t) = \begin{cases} \mathrm{PI}(t) & \text{for unsaturated period} \\ f_v(t - \Delta t_i') & \text{for saturated period.} \end{cases} \tag{2}$$

According to (1) and (2), the real infiltration rate $f_i(t)$ is determined by the parallel shift of the infiltration capacity f_v along the time axis (Fig. 1). To calculate $f(t)$, we must determine the ponding time t_s and the time shift $\Delta t' = t_s - t_0$. In accordance with our basic assumption there hold (MLS, [20]; PESCHKE [22]):

$$\int_0^{t_s} \mathrm{PI}(\tau)\,\mathrm{d}\tau = \int_0^{t_0} f_v(\tau)\,\mathrm{d}\tau \tag{3}$$

and

$$\mathrm{PI}(t_s) = f_v(t_0)\,. \tag{4}$$

In pracitcal hydrology the precipitation histogram $\mathrm{PI}(t)$ is given mainly as a discrete function of time (see also Fig. 2):

$$\mathrm{PI}(t) = \mathrm{PI}_i \quad \text{for} \quad (i - 1)\cdot\Delta t \leqq t < i\cdot\Delta t \tag{5}$$

and

$$i = 1, 2, \ldots, n\,.$$

Then, the conditional equations (3) and (4) read as follows:

$$S_{m-1} + [t_s - (m-1)\,\Delta t]\,\mathrm{PI}_m = \int_0^{t_0} f_v(\tau)\,\mathrm{d}\tau \tag{6}$$

and

$$\mathrm{PI}(t_s) = \mathrm{PI}_m = f_v(t_0)\,. \tag{7}$$

$S_{m-1} = \Delta t \cdot \sum_{i=1}^{m-1} \mathrm{PI}_i$, and m designates the time interval in which the soil surface becomes saturated. From the equations (6) and (7) the

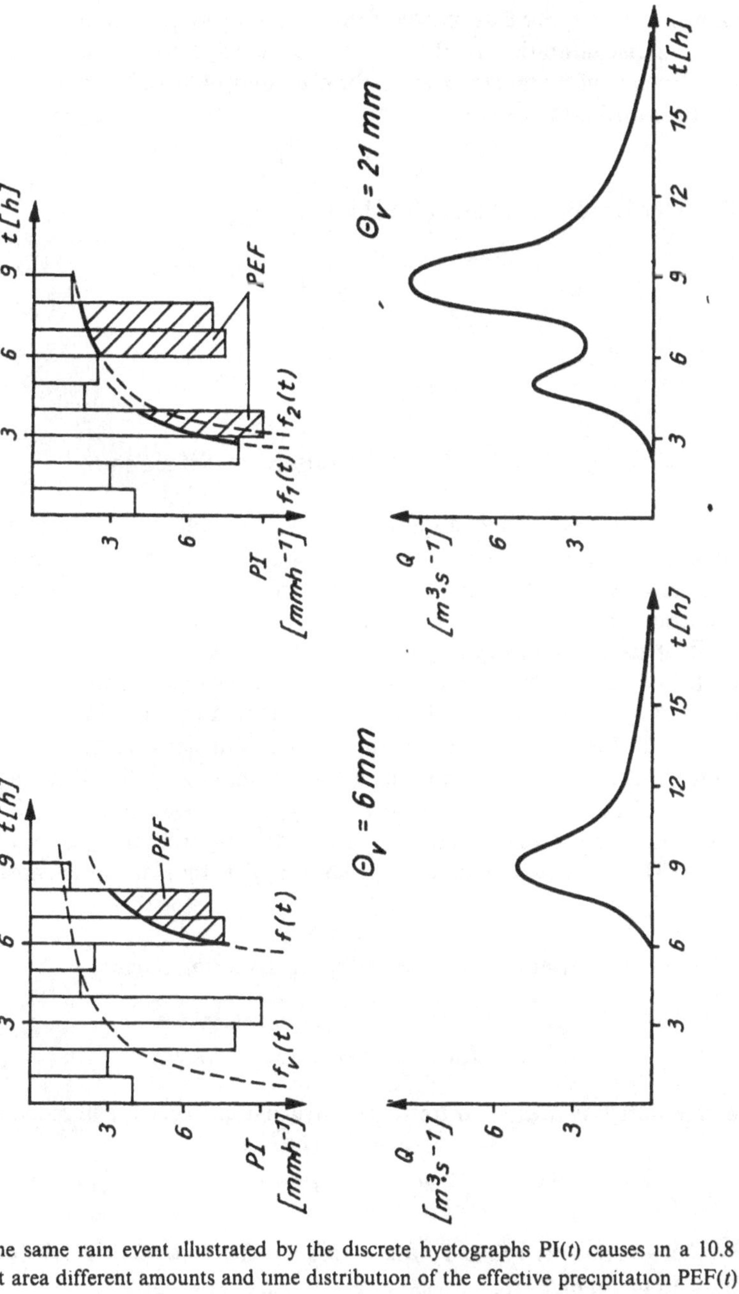

Fig. 2. The same rain event illustrated by the discrete hyetographs PI(t) causes in a 10.8 km^2 catchment area different amounts and time distribution of the effective precipitation PEF(t) and river discharge $Q(t)$ in dependence on variable antecedent soil moisture Θ_v

unknowns t_s and t_0 are determined if the instant of saturation lies within the mth interval. If the saturation of the soil surface is achieved in the moment of the sudden increase of the rain intensity at the beginning of the mth interval, the ponding time will be given immediately:

$$t_s = (m - 1)\,\Delta t\,. \tag{8}$$

With (8) the equations (6) and (7) reduce to

$$S_{m-1} = \int_0^{t_0} f_v(\tau)\,\mathrm{d}\tau \tag{9}$$

and

$$\mathrm{PI}_m > f_v(t_0)\,. \tag{10}$$

6. Consideration of the antecedent soil moisture

If infiltration tests are carried out at different antecedent soil moistures and the parameters of the applied empirical infiltration equation are determined from the results, we will find that, in general, these parameters are dependent on moisture. In the strict sense, of course, this is also true for the increase of moisture during the rain. This moisture dependence, however, is taken into account by the time shift. If we want to remain true to our model conception, we must apply the principle of time shift also to the antecedent soil moisture. This means that the hitherto used function $f_v(t)$ with its parameters is assigned to the dry soil. If the soil profile contains antecedent soil moisture amounting to Θ_v already at the beginning of rain, the infiltration capacity $f_v(\Theta_v)$ – again because of our basic assumption – is reduced as compared with the infiltration capacity of the dry soil $f_v(t)$. This reduction is effected by shifting $f_v(t)$ by t_v towards shorter times,

$$f_v(\Theta_v) = f_v(t + t_v) \tag{11}$$

(cp. Fig. 1). Instead of (1), now the following definition holds:

$$f(t) = \begin{cases} \mathrm{PI}(t) & \text{for } t \leqq t_s \\ f_v(t - \Delta t + t_v) & \text{for } t > t_s\,. \end{cases} \tag{12}$$

The time shift t_v is calculated from the particular antecedent soil moisture given:

$$\Theta_v = \int_0^{t_v} f_v(\tau)\cdot\mathrm{d}\tau\,. \tag{13}$$

The other unknowns, t_s and t_v, are obtained when equ. (11) is substituted into the equations (6) and (7).

7. The BLOINF model

With a simple time function of the infiltration capacity, given by an empirical infiltration equation, the above equations can be solved. We demonstrate this taking into account the following expressions for the infiltration capacity:

a) $f_v(t) = a\,t^{-\alpha}$ (Kostjakov [13])
b) $f_v(t) = a\,t^{-\alpha} + b$ (Mezencev [19])
c) $f_v(t) = a\,t^{-1/2} + b$ (Philip [27])
d) $f_v(t) = (f_0 - f_c)\,e^{-kt} + f_c$ (Horton [12])

where a, b, α, f_0, f_c and k designate empirical parameters. With the inclusion of these equations the slgorithm of the infiltration model BLOINF can be respresented as follows:

1. First of all, we determine the time shift t_v corresponding to the antecedent soil moisture. From equ. (13) there follows:

$$\text{a)}\quad t_v = \left[\frac{\Theta_v \cdot (1-\alpha)}{a}\right]^{\frac{1}{1-\alpha}}$$

$$\text{b)}\quad t_v + \frac{a}{b\cdot(1-\alpha)}\,t_v^{1-\alpha} - \frac{\Theta_v}{b} = 0$$

$$\text{c)}\quad t_v = b^{-2}\cdot(\Theta_v b + 2a\,[a(\Theta_v b + a^2)^{1/2}]) \tag{14}$$

$$\text{d)}\quad t_v + \frac{f_0 - f_c}{f_c \cdot k}\cdot(1 - e^{-kt_v}) - \frac{\Theta_v}{f_c} = 0$$

2. Step by step (for $m = 1, 2, \ldots$) we check the fulfilling of the test

$$\text{a)}\quad S_m > \frac{a}{1-\alpha}\left(\frac{a}{\mathrm{PI}_m}\right)^{\frac{1}{\alpha}-1} - \Theta_v$$

$$\text{b)}\quad S_m > \frac{a}{1-\alpha}\left(\frac{a}{\mathrm{PI}_m - b}\right)^{\frac{1}{\alpha}-1} + b\left(\frac{a}{\mathrm{PI}_m - b}\right)^{\frac{1}{\alpha}} - \Theta_v \tag{15}$$

$$\text{c)}\quad S_m > \frac{2a^2}{\mathrm{PI}_m - b} + b\left(\frac{a}{\mathrm{PI}_m - b}\right)^{\frac{2}{1}} - \Theta_v$$

$$\text{d)}\quad S_m > \frac{1}{k}\left[f_0 - \mathrm{PI}_m - f_c \ln\frac{\mathrm{PI}_m - f_c}{f_0 - f_c}\right] - \Theta_v$$

By this the interval of precipitation m is determined in which the soil surface becomes saturated for the first time.

3. First, we assume that the saturation occurs at the beginning of the mth interval, i.e.

$$t_s = (m - 1)\,\Delta t \tag{8}$$

then we calculate t_0 from (6) by using (8) and (11), and we get

$$\begin{aligned}
&\text{a)}\; t_0 = \left[\frac{1-\alpha}{a}\,(S_{m-1} + \Theta_v)\right]^{\frac{1}{1-\alpha}} - t_v \\
&\text{b)}\; t_0 + t_v + \frac{a}{b(1-\alpha)}\,(t_0 + t_v)^{1-\alpha} - \frac{S_{m-1} + \Theta_v}{b} = 0 \\
&\text{c)}\; t_0 = b^{-2}((S_{m-1} + \Theta_v)\,b + 2a[a((S_{m-1} + \Theta_v)\,b + a^2)^{1/2}]) - t_v \\
&\text{d)}\; t_0 + t_v + \frac{f_0 - f_c}{f_c \cdot k}\,(1 - e^{-k(t_0 + t_v)}) - \frac{S_{m-1} + \Theta_v}{f_c} = 0\,.
\end{aligned} \tag{16}$$

4. Now, the fulfilling of the tst is checked:

$$\begin{aligned}
&\text{a)}\; t_0 > T = \left(\frac{a}{\mathrm{PI}_m}\right)^{1/\alpha} - t_v \\
&\text{b)}\; t_0 > T = \left(\frac{a}{\mathrm{PI}_m - b}\right)^{1/\alpha} - t_v \\
&\text{c)}\; t_0 > T = \left(\frac{a}{\mathrm{PI}_m - b}\right)^{2} - t_v \\
&\text{d)}\; t_0 > T = \frac{1}{k}\ln\frac{f_0 - f_c}{\mathrm{PI}_m - f_c} - t_v
\end{aligned} \tag{17}$$

In the case of a positive answer we calculate

$$\Delta t' = t_s - t_0 \tag{18}$$

and

$$\begin{aligned}
&\text{a)}\; f(t) = a(t - \Delta t' + t_v)^{-\alpha} \\
&\text{b)}\; f(t) = a(t - \Delta t' + t_v)^{-\alpha} + b \\
&\text{c)}\; f(t) = a(t - \Delta t' + t_v)^{-1/2} + b \\
&\text{d)}\; f(t) = (f_0 - f_c)\,e^{-k(t - \Delta t' + t_v)} + f_c\,.
\end{aligned} \tag{19}$$

5. If the answer is negative, the saturation does not occur at the beginning, but within the mth interval. In this case, we set $t_0 = T$ and determine t_s from

$$\text{a)}\ t_s = \frac{a}{\mathrm{PI}_m(1-\alpha)}\left(\frac{a}{\mathrm{PI}_m}\right)^{\frac{1}{\alpha}-1} - \frac{S_{m-1}+\Theta_v}{\mathrm{PI}_m} + (m-1)\,\Delta t$$

$$\text{b)}\ t_s = \frac{a}{\mathrm{PI}_m(1-\alpha)}\left(\frac{a}{\mathrm{PI}_m-b}\right)^{\frac{1}{\alpha}-1} + \frac{b}{\mathrm{PI}_m}\left(\frac{a}{\mathrm{PI}_m-b}\right)^{1/\alpha} + \frac{S_{m-1}+\Theta_v}{\mathrm{PI}_m} + (m-1)\,\Delta t \tag{20}$$

$$\text{c)}\ t_s = \frac{2a^2}{\mathrm{PI}_m(\mathrm{PI}_m-b)} + \frac{b}{\mathrm{PI}_m}\left(\frac{a}{\mathrm{PI}_m-b}\right)^2 - \frac{S_{m-1}+\Theta_v}{\mathrm{PI}_m} + (m-1)\,\Delta t$$

$$\text{d)}\ t_s = \frac{1}{k\mathrm{PI}_m}\left[f_0 - \mathrm{PI}_m - f_c \ln\left(\frac{\mathrm{PI}_m - f_c}{f_0 - f_c}\right)\right] - \frac{S_{m-1}+\Theta_v}{\mathrm{PI}_m} + (m-1)\,\Delta t$$

and thence, Δt and $f(t)$.

By this, the infiltration rate $f(t)$ for the first saturated period of the precipitation event is determined, with which we are able to separate the precipitation excess for the hydrologic application. If another succession of unsaturated and saturated periods follows, this algorithm has to be applied repeatedly.

In accordance with the 2nd conclusion in the section 4, with respect to accuracy none of the empirical infiltration equations should be preferred. Therefore, we decide in favour of the simplest of them, i.e. for the Kostjakov equation (a). With it the algorithm given by the expression (14) to (20) gets its simplest form. For this reason, in the common version of the BLOINF model the Kostjakov equation is used.

8. Application

For the hydrologic use to separate the precipitation excess the BLOINF model has been applied so far to hydrologic basins which are different in size (4.6 to 360 km^2). The Figure 2 shows an example from which especially the influence of the antecedent soil moisture Θ_v on the amount of precipitation excess PEF (shaded areas) and discharge Q is evident. A hypothetic event of heavy rain occurs in the same hydrologic basin, but at a different antecedent soil moisture. The respective discharge hydrographs are obtained if the precipitation excess calculated with BLOINF is convoluted with the unit hydrograph of the hydrologic basin. In contrast, Figures 3 and 4 show measured rain histograms and discharge hydrographs and compare the latter with those calculated from BLOINF and the unit hydrograph.

Since a and α are empirical parameters without any physically founded reference

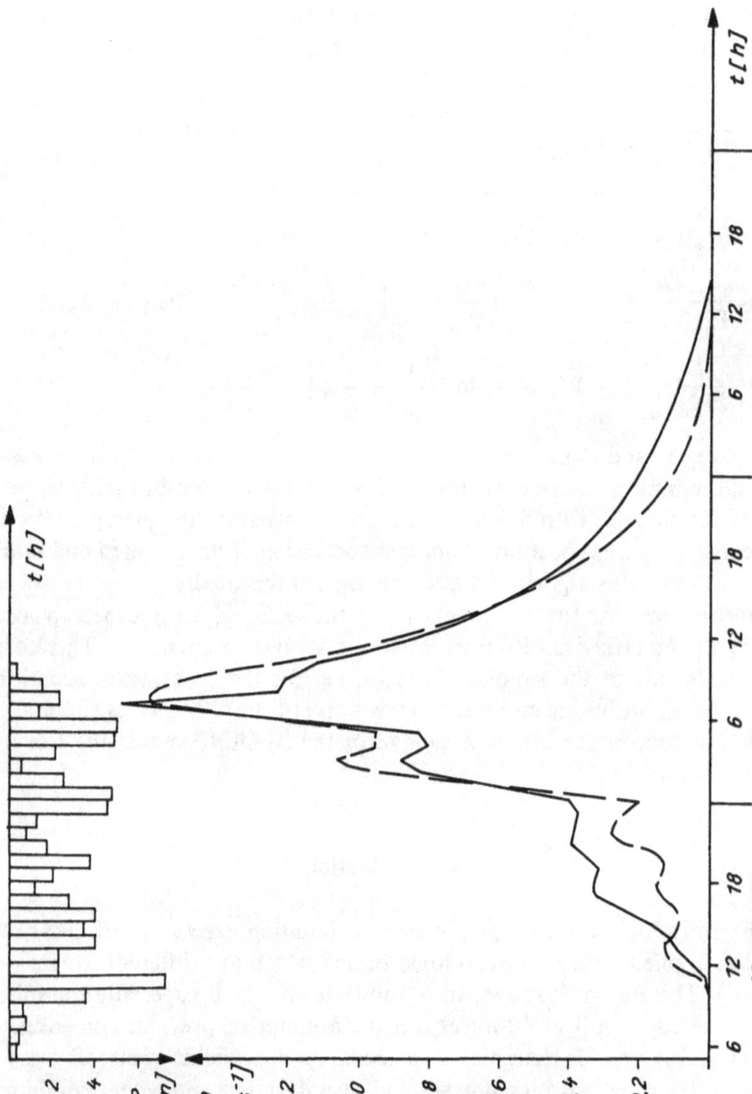

Fig. 3 Comparison between measured (———) and calculated (- - - - -) flood hydrographs of a small completely forested test area (4 6 km^2) in a hilly region The parameters used in the BLOINF model are α = 0 76, a = 13 6 mm h$^{\alpha-1}$ and Θ_v = 63 mm

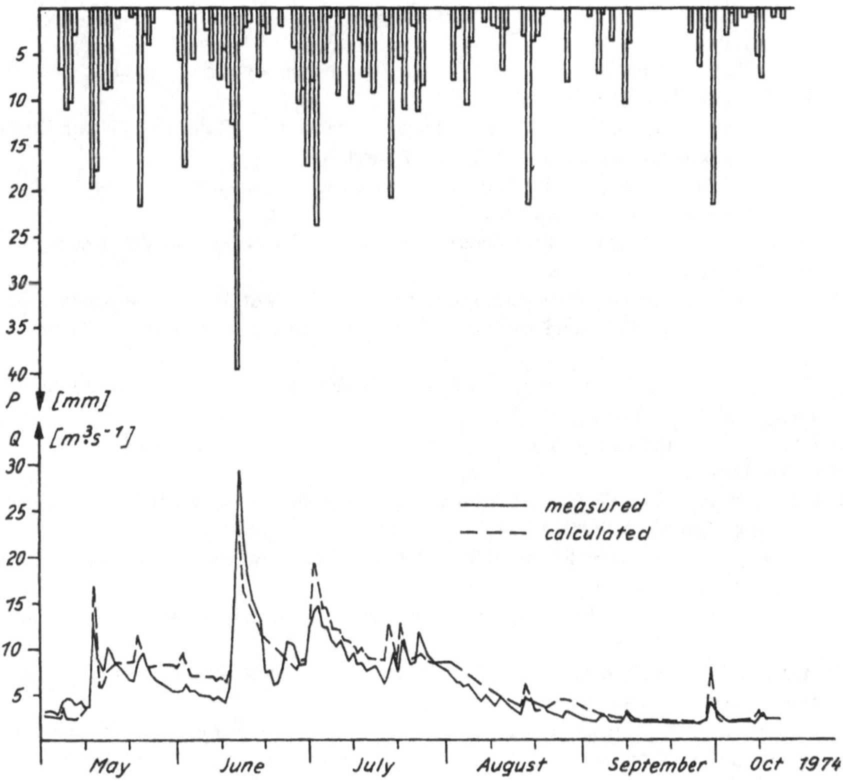

Fig. 4. Simulation of river discharge in a 363 km² basin using BLOINF within a complex watershed model

to the process of infiltration, simple optimization techniques can be used for determining them. As starting ranges there can be recommended for hydrologic basins of the low mountain ranges: $0.65 \leqq \alpha \leqq 0.85$ and $6 < \alpha < 14$, if PI and f_v are measured in mm $\cdot$ h^{-1}. Dyck and Peschke [6] report on a simple graphical application of BLOINF. If infiltration and runoff formation are very area-variable in larger hydrologic basins, a horizontal subdivision of the particular hydrologic basin has to be made. In such cases, the simplicity of BLOINF proves to be favourable. The determination of the parameters is simple, however, only if data of discharges measured for partial areas are available. If a calibration is not possible, an infiltration approach with physically defined parameters should be used (Peschke [23]).

References

[1] Childs, E. C. An introduction to the physical basis of soil water phenomena. – London/New York · J Wiley & Sons, 1969

[2] Corradini, C., Singh, V P Effect of spatical variability of effective rainfall on direct runoff by a geomorphologic approach. – J Hydrol **81** (1985) 1/2, 27–43

[3] Dawdy, D R, O'Donnell, T Mathematical models of catchment behaviour – Proc ASCE, HY4, paper 4410, J Hydraul Division **91** (1965), 123

[4] Dubreuil, P L. Review of field observations of runoff generation in the tropics. – J Hydrol. **80** (1985) 3/4, 237–264

[5] Dyck, S., et al Angewandte Hydrologie, Teil 2. – Berlin VEB Verlag f Bauwesen, 1978.

[6] Dyck, S., Peschke, G Gundlagen der Hydrologie. – Berlin VEB Verlag f Bauwesen, 1983

[7] Fedorov, S F Eksperimental'noe izučenie infil'tracii na slabopodzolistych počvach. – Trudy GGI **46** (100) (1954), 48–72.

[8] Fok, Y S Infiltration equation in exponential forms. – J Irrig and Drainage Div IR4 (1967), 125–135

[9] Green, W H, Ampt, G A. Studies on soil physics I The flow of air and water through soils. – J Agr Sci. **4** (1911), 1–24

[10] Holtan, H N A concept for infiltration estimates in watershed engineering. – U.S. Dept Agr, Agr Res. Service Pub (1961), 41–51

[11] Holtan, H N, Lopez, N C. USDAHL-70 model of watershed hydrology – USDA Techn. Bull. No. 1435 (1971).

[12] Horton, R. E. An approach toward a physical interpretation of infiltration capacity – Soil Sci Soc. Am. Proc. **5** (1940), 399–417

[13] Kostjakov, A N O dinamike koefficienta prosačivanija vody v počvo-grunty i neobchodimosti dinamičeskogo podchoda k ego izučeniju v meliorativnych celjach. – Počvovedenie **3** (1932), 293–297

[14] Kutergin, V A. K voprosu izučenija dinamiki vodoproicaemosti počv – Doklady Vaschnil (1939), vyp. 5–6.

[15] Kutilek, M Constant-rainfall infiltration. – J Hydrol **45** (1980), 289–303

[16] Lewis, M R. The rate of infiltration of water in irrigation practice. – Trans. Am. Geophys. Union **18** (1937), 361–368.

[17] Lichty, R. W, Dawdy, D R., Bergmann, J M Rainfall runoff model for small basin flood hydrograph simulation – Proc. Symp. The use of analog and digital computers in hydrology, IAHS/UNESCO, Tucson, Vol 2 (1968), 356–367

[18] Lin, W, Gray, D M. Physical simulation of infiltration equations. – Water Resources Res. **7** (1972) 5, 1234–1240

[19] Mezencev, V S. K teorii formirovanija poverchnostogno stoka so sklonov – Meteor i gidrol. **3** (1948), 33–40.

[20] Mls, J Effective rainfall estimation. – J Hydrol **45** (1980), 305–311

[21] Moore, I D, Larson, C. L. Hydrologic impact of draining small despressional watersheds. – J Irrig. and Drainage Div IR4 (1980), 345–363

[22] Peschke, G Infiltration unter einfachen Randbedingungen – Gerlands Beitr Geophys. **89** (1980) 1, 15–26.

[23] PESCHKE, G.: Soil moisture and runoff components from a physically founded approach. — Acta hydrophys. **31** (1987) 3/4, 191—205.

[24] PESCHKE, G.; MÜLLER, A.: Investigations on the infiltration process and comparison of the results obtained with a mathematical model. — J. Hydrol. Sci. **3** (1976) 3/4, 105—121.

[25] PESCHKE, G.; KUTILEK, M.: Infiltration model in simulated hydrographs. — J. Hydrol. **56** (1982) 3/4, 369—379.

[26] PESCHKE, G.; DUNGER, V.; GURTZ, J.: Changing of soil moisture by infiltration and evapotranspiration. — Budapest: IAHS — 2nd Scientific Assembly, July 1986, (accepted for publ. in the IAHS Publ.).

[27] PHILIP, J. R.: The theory of infiltration: 1. The infiltration equation and its solution. — Soil Sci. **83** (1957), 345—357.

[28] PHILIP, J. R.: The theory of infiltration: 4. Sorptivity and algebraic infiltration equations. — Soil Sci. **84** (1957), 257—264.

[29] POPOV, E. G.: Osnovy gidrologičeskich prognozov. — Leningrad: Gidrometeoizdat, 1968.

[30] SIEGERT, K.: Oberflächenabfluß von landwirtschaftlichen Nutzflächen infolge von Starkregen. — Mitt. Leichtweiss-Inst. f. Wasserb., TU Braunschweig, Heft **58** (1978).

[31] SKAGGS, R. W.; HUGGINS, L. F.; MONKE, E. J.; FOSTER, G. R.: Experimental evaluation of infiltration equations. — Trans. Am. Soc. Agr. Eng. **12** (1969) 6, 822—828.

[32] SMITH, R. E.: The infiltration envelope: Results from a theoretical infiltrometer. — J. Hydrol. **17** (1972) 1/2, 1—21.

[33] SMITH, R. E.; CHERRY, D. L.: Rainfall excess model from soil water flow theory. — J. Hydraul. Div. HY9 (1973), 1337—1351.

[34] SOZYKIN, N. F.: O dinamike vpityvanija vody v počvu. — Sb. „Maksimalnyj stok malych bassenov". — Moskva: Transželdorizdat, 1940.

[35] SWARTZENDRUBER, D.: Infiltration of constant-flux rainfall into soil as analyzed by the approach of Green and Ampt. — Soil Sci. **117** (1974) 5, 272—281.

[36] SWARTZENDRUBER, D.; HUBERTY, M. R.: Use of infiltration equation parameters to evaluate infiltration differences in the field. — Trans. Am. Geophys. Union **39** (1958), 84—93.

[37] SWARTZENDRUBER, D.; YOUNGS, E. G.: A comparison of physically-based infiltration equations. — Soil Sci. **117** (1974), 165—167.

[38] THEILE, K.: Ein Infiltrationsmodell unter Berücksichtigung des Landoberflächenabflusses. — Wasserw. — Wassert. **22** (1972), 415—419.

[39] TISDALL, A. L.; Infiltration studies on a red-brown earth in the Reverina. — Austral. Inst. Agr. Sci. **16** (1950), 26—29.

Acta hydrophys., Berlin 31 (1987) 3/4, S. 225–238

The Influence on Secchi Depth of Suspended Matter, Phytoplankton Composition, and Water Colour

By G. SCHELLENBERGER and R. STELLMACHER[1])

Summary On the base of daily measurements of SECCHI depth and relevant substances in the shallow eutrophic lake Müggelsee relationships between these parameters were set up to estimate the concentration of suspended matter and algal biomass from observations of the transparency According to these results the SECCHI depth seems not to depend on the algal composition. Apparent differences of the partial optical efficiencies described by some workers are an artefact of regression analysis in the case of disregarding the detritus as a quantity

Zusammenfassung Aus täglichen Messungen in einem eutrophen See wurden Regressionsgleichungen zur Abschätzung des Sestongehaltes und der Phytoplanktonbiomasse aus Sichttiefenbeobachtungen abgeleitet. Es zeigte sich, daß bei Berücksichtigung des Detritus die optische Effizienz der untersuchten Algengruppen praktisch gleich ist und – bei der Korrelation von Sichttiefen nur mit Algenkonzentrationen – festgestellte Unterschiede den abgestorbenen Algen zuzuschreiben sind, die anscheinend wegen artspezifischer Sedimentationsgeschwindigkeiten in unterschiedlicher Menge mit der jeweiligen Lebendkomponente gekoppelt sind.

Резюме. Были установлены уравнения регрессии для оценки содержания сестона и фитопланктона из наблюдений прозрачности воды по диску СЕККИ. Оказалось, что — при учете детрита — оптическая эффективность всех исследованных групп фитопланктона почти равной. Различия, определенные при корреляции прозрачности только с концентрациями групп водорослей, объясняются разными из-за различных скоростей седиментации долями „сопуствующих" мертвых особей.

1. General relations

As a simple but rough method to characterize the optical properties of natural waters the determination of transparency by means of the SECCHI depth is still in

[1]) Dr rer nat. GÜNTER SCHELLENBERGER, Dr sc. nat. RITA STELLMACHER, Akademie der Wissenschaften der DDR, Institut für Geographie und Geoökologie, Bereich Hydrologie, Müggelseedamm 260, Berlin, DDR-1162.

use. It consists in lowering into the water a white plate, noting the depth at which the plate disappears, then lifting the plate and noting the depth at which it reappears. The average of these two readings is termed SECCHI depth SD [12], [18]. According to the transparencies found in the investigated lakes (0.4 to 4 m) white disks of 25 cm diameter have been employed. Because the observation is disturbed f.i. by the sun-dependent and fluctuating reflection of light from the water surface, the determinations must be made through a sun- and wave-shaded area.

For the perception of an object in water it is necessary that the difference in brightness between the disk and its surrounding does not fall below the threshold of the human eye on the way from the disk to the observer. Fundamental contributions to the visibility in water were given by DUNTLEY [6], [14]. According to these general equations the reciprocal SECCHI depth is approximately a linear function of the sum $c + k$. c is the attenuation coefficient (physical extinction coefficient), k the downward radiance attenuation coefficient (vertical extinction coefficient). The constant value of the product $(c + k) \cdot \mathrm{SD}$ equals 8.69 according to TYLER and 8.98 according to HOLMES, respectively [1].

Measurements of the ratio k/c performed by us in shallow lakes resulted in values between 0.2 and 0.4 [19]. If the hydrooptical conditions are similar, relatively constant values of the products $c \cdot \mathrm{SD}$ and $k \cdot \mathrm{SD}$ can be anticipated, too. In a medium with a high scatterance but a low absorptance the ratio is small. It increases with rising absorbing properties of the natural water. Highly turbid eutrophic lakes are strongly scattering (milky) media. The portion of scatterance in the attenuation coefficient amounts to maximal values if the detritus-phytoplankton ratio is high [2]. The other extreme are dark-coloured waters poor in suspended particles. Therefore, the light intensity at SECCHI depth in different lakes with differing optical proporties can be highly variable.

As the relation of transparency to vertical extinction coefficient can be taken into account for the evaluation of primary productivity, this relationship has been established more than once. For instance, DOKULIL compiled values of the product $\mathrm{SD} \cdot k$ found by several workers [5]. Including 5 own measurements quantities ranging from 1.0 to 3.64 were obtained (mean value: 2.23, median: 2.08). Another compilation of data from Byelorussian lakes was derived by IKONNIKOV [13]. He found the following mean relationship:

$$k \cdot \mathrm{SD} = (2.09 \pm 0.06) + 0.02 \cdot \mathrm{SD},$$

which fairly corresponds to the results by DOKULIL. Because the vertical attenuation of the downward irradiance E_d can be approximated by the exponential function

$$E_d(z) = E_d(0)\, e^{-k \cdot z},$$

the light intensity at the SECCHI depth z = SD is ranging from 37.8% of the subsurface light $E_d(0)$ for $k \cdot \text{SD} = 1$ to 2.6% for $k \cdot \text{SD} = 3.64$.

Many attempts have been undertaken to use the SECCHI disk depth for a rough evaluation of suspended matter (seston) or the concentration of phytoplankton. In the latter case as a measure of the algal content the chlorophyll concentration has been more frequently used. Although a general relationship between the SECCHI depth and the attenuation coefficient exists, the dependence on the seston concentration must be weaker, because the attenuation of light is related to the whole cross-section area of the involved particles and not to the volume or the weight of suspendoids. This means that the factor of proportionality between the seston content G and the attenuation coefficient of particles c^p contains the product $\varrho \cdot d$, in which ϱ is the density and d the mean diameter of the particles involved [10]. For all this we could point out that the approximative equation established by us was fairly valid in lakes near Berlin [9]:

$$G(\text{mg/l dry weight}) = 7 \cdot c^p_{588}\ (\text{lg/m})\,.$$

The attenuation coefficient of particles based on decade logarithms has been measured at the wave length of 588 nm.

The composition and with it the optical properties of seston show annual variations. Maximal summer concentrations contain a higher portion of algae than minimal quantities in winter. The absorptance by dissolved organic constituents is less detectable in the presence of many particles. As a result of these facts many investigators found a depressed optical efficiency at high seston concentrations: According to BUL'ON 1977, see [13])

$$\text{SD} \cdot G^{0.70} = 4.0\,,$$

according to IKONNIKOV (1979)

$$\text{SD} \cdot G^{(0.67 \pm 0.09)} = 4.62 \pm 1.2\,, \quad n = 134,\ r^2 = 0.33\,, \quad 0.7 < G < 42 \text{ mg/l dry weight},$$

according to DOKULIL (1979)

$$\text{SD} \cdot G^{0.83} = 5.91\,, \qquad n = 160, \qquad r^2 = 0.69$$

(for the highly turbid and scattering water of lake Neusiedler See DOKULIL stated a diminished constant, minor to the bulk of the other lakes

$$\text{SD} \cdot G^{0.61} = 2.10, \qquad n = 87, \qquad r^2 = 0.41)\,.$$

The number of sampling units for each regression is n, and r^2 is the proportion of sum of squares attributable to the regression.

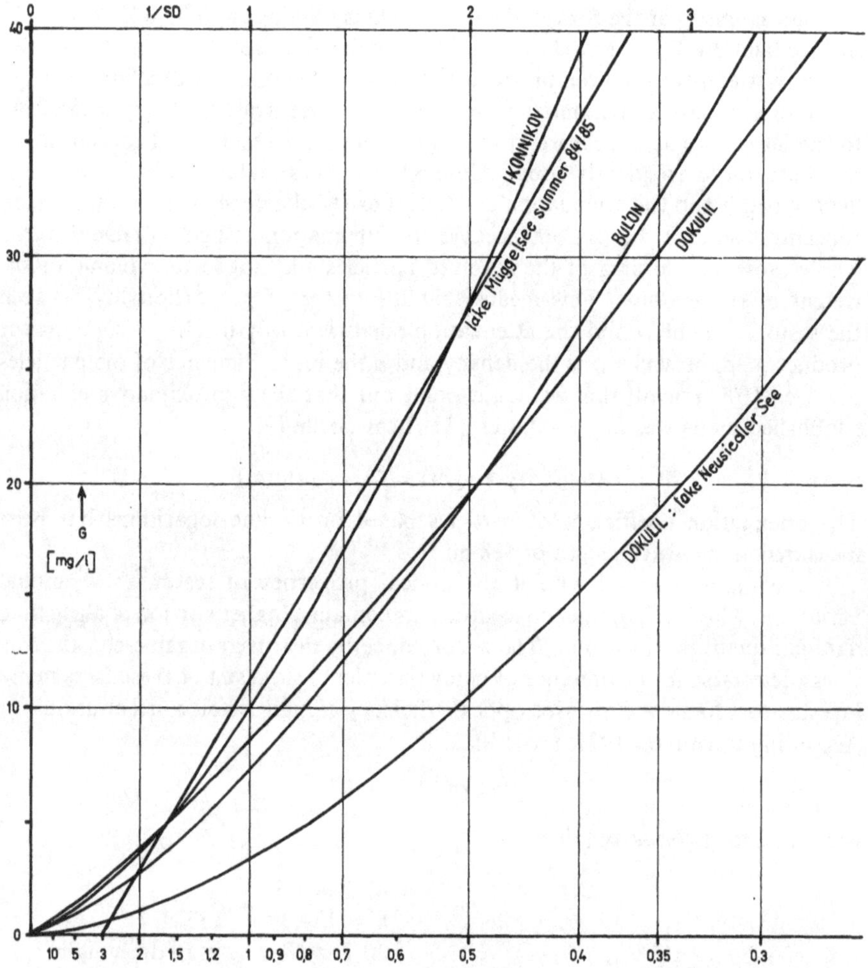

Fig. 1. Relationship between seston concentration G and SECCHI depth SD according to the compilations of some workers

The spread of the lines in Fig. 1 illustrates that in spite of apparently similar coefficients significant differences between the determined regressions occur. (In this paper we do not include those dependences not basing on the principal relationship $1/SD \sim c$ [8], [22].)

Concerning the relationship between transparency (m) and chlorophyll-a concentration (mg/m^3) many authors found also an exponent <1:

KUZ'MICEVA (1970, see [13])

$$SD \cdot Chl^{0.60} = 7.21, \qquad 10 < Chl < 200;$$

BUL'ON (1977, see [13])

$$SD \cdot Chl^{0.46 \pm 0.04} = 6.46 \pm 0.62, \qquad n = 109, \qquad Chl < 150;$$

CARLSON (1977)

$$SD \cdot Chl^{0.68} = 7.59, \qquad \text{data from 78 lakes;}$$

IKONNIKOV (1979)

$$SD \cdot Chl^{0.56 \pm 0.10} = 7.05 \pm 1.19, \qquad n = 165, \qquad r^2 = 0.23, \quad 1 < Chl < 180;$$

FORSBERG/RYDING (1980)

$$SD \cdot Chl^{0.57} = 7.08, \qquad 4 < Chl < 200;$$

CARLSON (1980)

$$1/SD = a + b\, Chl,$$

the range of the parameters a, b is as follows:

$$0.034 < a < 0.755, \quad \text{median value} \quad a_{1/2} = 0.373,$$
$$0.0075 < b < 0.0536, \quad b_{1/2} = 0.0182.$$

LOTH (1985)

$$1/SD = 0.019 + 0.0395\, Chl$$

and

$$1/SD = 0.028 + 0.200 \cdot B, \text{ respectively, } B < 3 \text{ mg/l fresh weight}$$

B: algal biomass.

HOSPER (1985)

data from lake Veluwe, mean values for Oscillatoria-dominated periods

$$1/SD = 0.016\, Chl, \text{ average 1977—79,}$$
$$1/SD = 0.029\, Chl, \text{ average 1980—83, after restoring remedies;}$$

BREZONIK [1] considers not only the dependence of transparency on the chlorophyll concentration but also on the water colour Pt and on the turbidity. Comparing BREZONIK's with our results we cite in this paper only the regression BREZONIK (1978)

$$1/SD = 0.36 + 0.03\, Chl + 0.001\, Pt, \qquad n = 55, \qquad r^2 = 0.62.$$

As expected the scattering between the particular regression lines is large. Confronting the margin curves from BUL'ON and CARLSON (1977) we notice that a SECCHI depth of 0.6 m corresponds to a chlorophyll concentration of 175 and 42 mg/m^3, respectively.

2. The relationship between Secchi depth and substances in the eutrophic lake Müggelsee

On the base of our daily measurements of SECCHI depth and relevant parameters in the shallow lake Müggelsee we examined the possibility to estimate seston and phytoplankton concentrations from transparency observations. For the years 1982—1984 we used 144 pentade means (5 consecutive days) of the following parameters:

- SECCHI depth,
- attenuation coefficient of particles c^P at 588 nm, as a measure of seston concentration,
- attenuation coefficient of dissolved substances c^F at 428 nm, as a measure of water colour,
- biomasses of the phytoplankton groups:
  + blue-greens (Cy),
  + diatoms, partitioned into *Stephanodiscus* (St) and other species (Do), because the present dominating species *Stephanodiscus astraea* shows deviating optical properties [21],
  + remaining species (Re).

We used pentade means to reduce random errors of the visually determined transparency. As the water depth at the observation site amounts to 2.3 m, the range of investigated SECCHI depths is limited by this value. For the observed period the distribution of the products SD $\cdot$ c^P (that means also for SD $\cdot$ G) nearly follows a normal (GAUSSIAN) distribution.

For the dependence $1/SD = f(c^P)$ we obtained the following linear regression:

$$1/SD = 0.307 + 0.433\, c^P, \quad \text{for} \quad 0.32 < SD < 1.59, \quad 0.26 < c^P < 2.87, \quad n = 144, \quad r^2 = 0.90. \tag{1}$$

Including the water colour c^F as an additional variable the relationship can be improved only a bit.

$$1/SD = 0.186 + 0.422\, c^P + 0.198\, c^F, \quad r^2 = 0.92\,. \tag{2}$$

Employing actual values of the water colour of lake Müggelsee this dependence is plotted in Fig. 2. For practical use the attenuation coefficient c^P was replaced

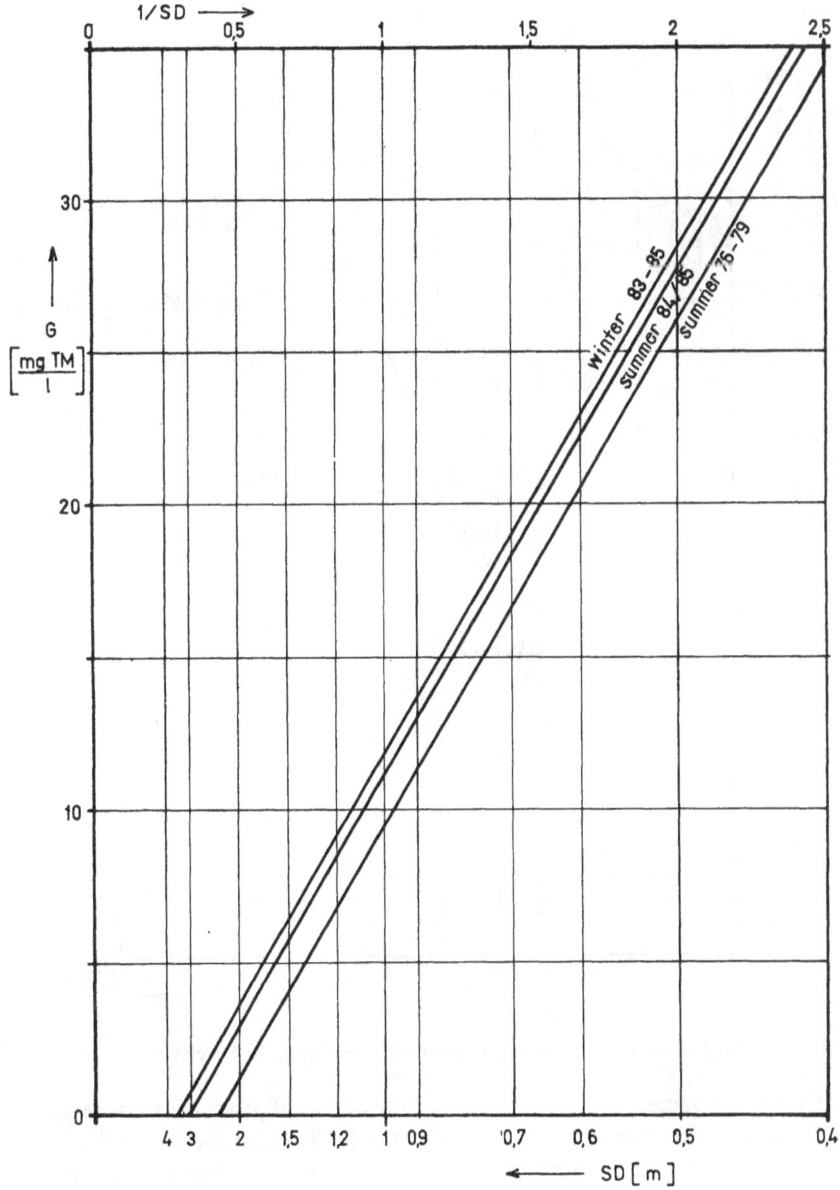

Fig. 2 Estimates of the seston content G from observations of the SECCHI depth SD (for application in the Berlin lake system)

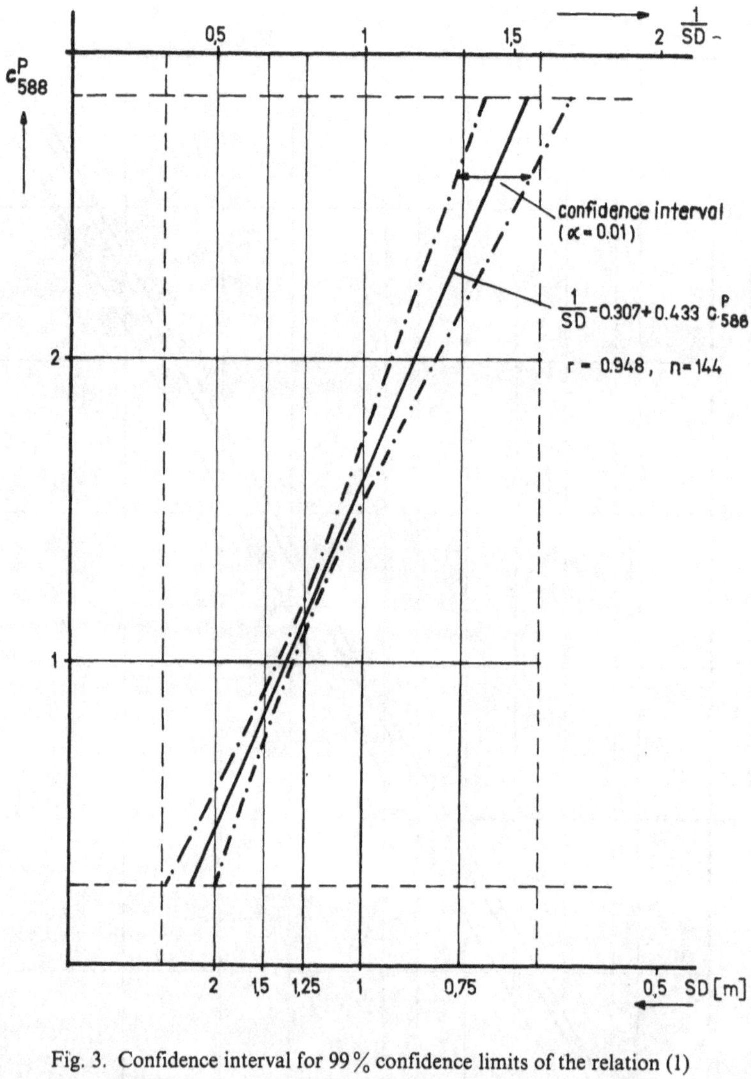

Fig. 3. Confidence interval for 99 % confidence limits of the relation (1)

by the seston concentration G. To estimate the degree of precision of the regression (1) the confidence interval for 99 % confidence limits is represented in Fig. 3.

Concerning the relationship between SECCHI depth and algal groups we found the following regression:

$$1/SD = 0.806 + 0.162\ Cy_s + 0.063\ St_s + 0.045\ Do_s + 0.101\ Re_s, \qquad r^2 = 0.56. \tag{3}$$

The subscript s means that standardized values have been used. The correlation coefficient of regression (3) is diminished in comparison with eqs. 2 and 3, because the relation (3) contains only the various algal groups but not explicitely the detritus. Its influence is hidden in an enlarged intercept and possibly in altered partial slopes. The slopes of the particular algal groups are different. For instance, the slope of the blue-green algae is three to four times the slope of diatoms. But from this fact we cannot necessarily conclude that different slopes of partial algal groups are the result of their differing optical efficiencies. This may be also an effect of just died algae with species-dependent sedimentation velocities.

The moderate relationship between seston and algae accounts for the diminished correlation coefficient of the regression (3) as we learn from the following equation:

$$c_s^p = 0.457\,\mathrm{Cy}_s + 0.274\,\mathrm{St}_s + 0.196\,\mathrm{Do}_s + 0.502\,\mathrm{Re}_s\,, \qquad r^2 = 0.57\,. \tag{4}$$

According to the former explanations we could expect a closer relationship if the detritus content would be involved.

As direct detritus measurements are not available, we include in eq. (5) the quantity B/c^p as a measure of the algal portion in seston:

$$1/\mathrm{SD} = 0.702 + 0.04\,\mathrm{Cy} + 0.032\,\mathrm{St} + 0.038\,\mathrm{Do} + 0.038\,\mathrm{Re} - 0.03\,\mathrm{B}/c^p \quad r^2 = 0.82\,. \tag{5}$$

With this regression a similarly strong correlation is obtained as with eq. (1). Further, eq. (5) shows the correctness of our assumption that the partial slopes of algal groups are changed if the detritus is taken into consideration. The partial slopes of eq. (5) are smaller than those of eq. (3) and have similar values.

3. Conclusion

The Secchi disk depth seems not to depend on the algal composition. Apparent differences of the partial optical efficiencies are an artefact of regression analysis in the case of disregarding the detritus as a quantity. Just died algae "accompanying" the living stock are "responsible" for apparently differing optical efficiencies.

Dividing the partial efficiencies of eq. (3) by those of eq. (5) we get an estimate of the species-dependent accompanying dead fraction.

Blue-greens: 0.162/0.040 = 4.05,
Stephanodiscus: 0.063/0.032 = 1.97,
other diatoms: 0.045/0.038 = 1.18,
remaining species: 0.101/0.038 = 2.66.

The slowly sedimentating detritus from blue-green algae results in a greater portion of dead organisms than in the case of diatoms.

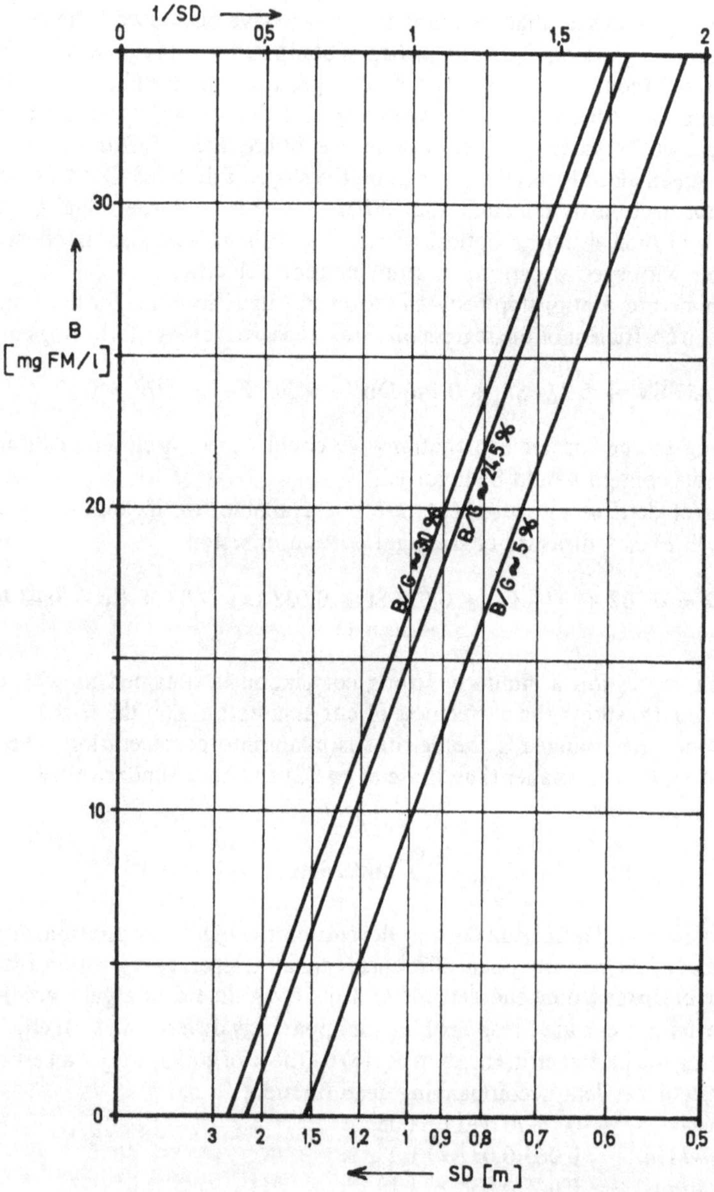

Fig. 4. Estimates of the algal biomass B from observations of the SECCHI depth SD for various biomass – seston ratios

Because the optical efficiencies of the partial algal groups are similar, it is no longer necessary to partition the whole phytoplankton biomass B into components

$$1/\mathrm{SD} = 0.702 + 0.0364\ \mathrm{B} - 0.03\ \mathrm{B}/c^{\mathrm{P}}\,. \tag{6}$$

With an increasing phytoplankton-seston ratio the intercept of the regression $1/\mathrm{SD} = \mathrm{f}(c^{\mathrm{P}})$ is diminishing. The regression (6) is plotted in Fig. 4. For practical applications in the Berlin lake system the variable $\mathrm{B}/c^{\mathrm{P}}$ was replaced by observed values: minimal values in winter (1.75), maximal values in summer (10.5), mean value for the vegetation period (April to October): 8.58.

Because in lake Müggelsee the mean chlorophyll-algal biomass ratio (Chl/B) amounts to $3.64^0/_{00}$ [20], the optical efficiency coefficient of chlorophyll-a seems to be 0.01. Comparing this value with the coefficients found by Brezonik (0.03), Carlson (0.0075 to 0.0536), Loth (0.0395) and Hosper (0.016; 0.029) we detect variable detritus portions by the degree of augmentation.

In this paper we do not discuss the dependence of the vertical extinction coefficient on the chlorophyll concentration, but we wish to emphasize that neither this problem can be solved disregarding the accompanying detritus. Therefore, the discussion between Carlson and his critics Lorenzen as well as Megard et al. seems not to be focussed on the main point [4], [15], [17].

For the dependence of the vertical extinction coefficient (spectral average) on substances in water the following general equation is valid:

$$k = a + b \cdot \mathrm{B} + c \cdot \mathrm{D}\,,$$

where a pertains to attenuance by the water itself and by dissolved organic substances (yellow substance).

D is the detritus concentration, which can be divided into two parts: the first one Df incorporates recently died algae with a synchronous course to the living component, the second one involves refractory organic compounds Dr not related to the actual density of phytoplankton [21].

$$\mathrm{D} = \mathrm{Df} + \mathrm{Dr} = \nu \cdot \mathrm{B} + \mathrm{Dr}\,.$$

Regression analysis considering only the dependence of k on B (or Chl, respectively) yields the following equation:

$$k = a + c \cdot \mathrm{Dr} + (b + \nu \cdot c)\,\mathrm{B}\,.$$

In this case we notice an enlarged intercept $a + c \cdot \mathrm{Dr}$ and also an enlarged slope $b + \nu \cdot c$.

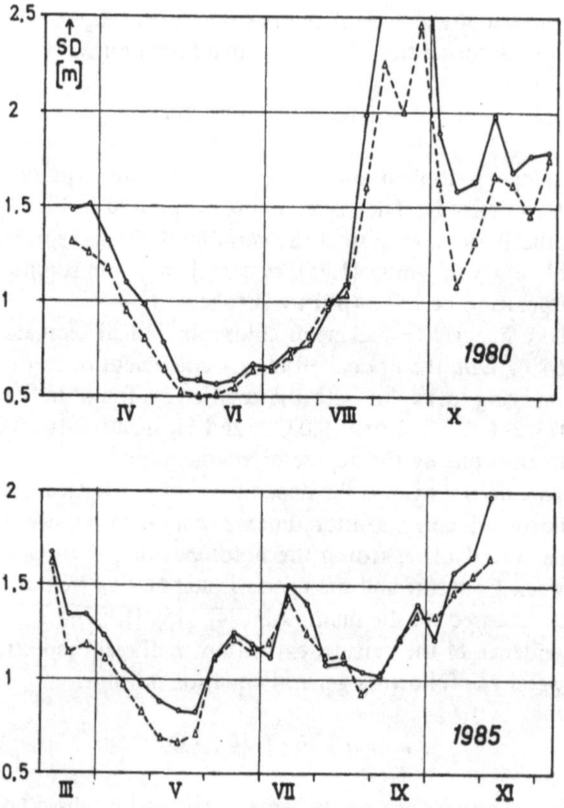

Fig. 5. Annual courses of SECCHI depth 1980 and 1985 in lake Müggelsee
Regression model (1) (broken lines) as compared with measurements (solid lines)

Similar relationships are valid for the reciprocal SECCHI depth. If you find a range of apparent optical efficiencies of chlorophyll or algal biomass, then the minimal value is disturbed by the accompanying fresh detritus least at all (see CARLSON [4], HOSPER [11]).

Fig. 5 illustrates the testing of the regression (1) by means of values for the years 1980 and 1985 not used in calibrating the model. On the whole the relationship (1) seems to fit the measurements fairly well, but there are also some discrepancies (the estimate for the period March to May is too low). Of course, we cannot overlook the fact that a precision of 0.1 m requires a strong visual acuity of the observer. Because the model was calibrated with SECCHI depths less than 2.5 m, it is not able to describe greater transparencies (see clear water phase in September 1980).

Acknowledgement

We are indebted to thank Mrs. S. HOEG for her contribution to this work by performing the algal analysis and most of the daily measurements.

References

[1] BREZONIK, P. L.: Effect of organic color and turbidity on Secchi disc transparency. — J. Fish. Res. Can. **35** (1978) 11, 1410—1416.

[2] CARDER, K. L.; STEWARD, R. G.: A remote-sensing reflectance model of a red-tide dinoflagelate off West Florida.— Limnol. Oceanogr. **30** (1985), 286—298.

[3] CARLSON, R. E.: A trophic state index for lakes.— Limnol. Oceanogr. **22** (1977), 361—369.

[4] CARLSON, R. E.: More complications in the chlorophyll — Secchi disk relationship. — Limnol. Oceanogr. **25** (1980), 379—382.

[5] DOKULIL, M.: Optical properties, colour, and turbidity. — In: Neusiedler See, ch. 15. — Den Haag: Junk Publ., 1979.

[6] DUNTLEY, S. O.: Light in the sea.— (Contr. Scripps Inst. Oceanogr. **33** (1963), No. 1499), J. Opt. Soc. Am. **53** (1963), 214—233.

[7] FORSBERG, C.; RYDING, S. O.: Eutrophication parameters and trophic state indices in 30 Swedish waste-receiving lakes. — Arch. Hydrobiol. **89** (1980), 189—207.

[8] GEORGI, F.: Untersuchungen zur Variablilität von Quantität und Qualität des Sestons in den Darß-Zingster Boddengewässern.— Rostock: WPU, Diss., 1983.

[9] HOEG, S.; SCHELLENBERGER, G.: Über Änderungen der Lichtextinktion in einem eutrophen See und ihre Ursachen. — Acta hydrophys. **13** (1968) 1, 11—60.

[10] HOEG, S.; SCHELLENBERGER, G.: Zur Messung der Sinkstoffkonzentration auf optischem Wege. — Acta hydrophys. **16** (1972) 3, 133—144.

[11] HOSPER, S. H.: Restoration of lake Veluwe, the Netherlands, by reduction of phosphorus loading and flushing. — Wat. Sci. Tech. **17** (1985), 757—768.

[12] HUTCHINSON, G. E.: A treatise on limnology. Vol 1, Geography, Physics, and Chemistry. — New York: 1957.

[13] IKONNIKOV, V. F.: Dependence of the light conditions in water on the content of chlorophyll and seston. (Russ.) — In: Obshchie osnovy izucheniya vodnykh ekosistem. — Leningrad: Nauka, 1979, 199—206.

[14] JERLOV, N. G.: Marine optics. Amsterdam, Oxford, New York: 1976.

[15] LORENZEN, M. W.: Use of chlorophyll-SECCHI disk relationship. — Limnol. Oceanogr. **25** (1980), 371—372.

[16] LOTH, P.: Zusammenhänge zwischen Chlorophyll a, Phytoplanktonvolumen, Seston, Sichttiefe und Trübung als Methoden zur Einschätzung der Wasserbeschaffenheit in Talsperren. — Acta hydrochim. hydrobiol. **13** (1985), 319—329.

[17] MEGARD, R. O.; et al.: Light, SECCHI disks, and trophic state. — Limnol. Oceanogr. **25** (1980), 373—377.

[18] RICHTER, D.; WIEDRICH, R.: Die Sichttiefen im Stechlin- und Nehmitzsee. — Limnologica **5** (1967), 299—305.

[19] SCHELLENBERGER, G.: Ellipsoidale Approximation der Strahlungsverteilung in Gewässern. — Acta hydrophys. **15** (1970) 1, 23—41.

[20] SCHELLENBERGER, G.; STELLMACHER, R.; HOEG, S.; ROHDE, E.: Zusammenhang zwischen Biomassen von einzelnenAlgengruppen, Chorophyllkonzentrationen und Floureszenzwerten im Müggelsee. — Acta hydrophys. **29** (1985) 2/3, 211—220.

[21] SCHELLENBERGER, G.; HOEG, S.: Seston, Phytoplankton und Wasserfarbe in zwei verbundenen Flachseen — Vergleich und zeitliche Veränderungen. — Acta hydrophys. **30** (1986) 4, 231—273.

[22] SCHINKOWSKI, H.: Untersuchung über den Einfluß einiger produktionsbiologischer Parameter auf die Sichttiefe im Meer. — Kieler Meeresforschungen **27** (1971), 4—19.

Acta hydrophys., Berlin 31 (1987) 3/4, S. 239–250

Zur Kontrolle der flächenhaften Gewässerbelastung infolge atmosphärischen Stoffeintrags und durch die Landwirtschaft mittels Drainagemeßfeldern

Von W. GERDS[1]) und N. HANSEL[2])

Zusammenfassung Die Kontrolle der flächenhaft bedingten Gewässerbelastung ist für die Wasserwirtschaft und die Humanökologie ein ernsthaftes Problem. Dabei ist die Gewässerbelastung durch die Landwirtschaft besonders wichtig. Für die Kontrolle der landwirtschaftlichen Verschmutzungsquellen wird als eine mögliche Methode der Betrieb von Drainagemeßfeldern empfohlen. Damit sind verschiedene Messungen und Beobachtungen an Drainagen oder ihren Teilen gemeint.

Anhand der Drainage Börtewitz wird die Methode gezeigt Das ist ein drainiertes Feld von 30 ha Größe in der zentralsächsischen Lößregion (210 bis 220 m ü NN).

Drainagemeßfelder können benutzt werden

1 als Teil eines örtlichen Sondermeßnetzes, z. B. in Trinkwasserschutzzonen,

2. in einem landesweiten Meßnetz,

3 als agrarökologisches Observatorium zu Forschungszwecken Für die Auswahl, die Einrichtung und den Betrieb von Drainagemeßfeldern werden Gesichtspunkte genannt.

Summary The monitoring of water pollution by non-point sources is a serious problem of water management and human-ecology In this connection the influences of agriculture are very important. As a possible method for monitoring agricultural sources of water pollution is proposed to practise drainage plots. That means different measurements and observations at a drainage network or at a part of it.

The method is shown by the pilot plot Börtewitz – a drained field of 30 ha (210 to 220 m above SL) in the Loess Region of Central Saxony

Drainage plots can be used

– firstly as part of local special network, e.g. in drinking water protection zones,

– secondly in a land wide measurement network,

– thirdly as agroecological like a research station.

Recommendations are given to select, install and operate drainage plots.

---

[1]) Dr rer nat. WOLFGANG GERDS, Akademie der Wissenschaften der DDR, Institut für Geographie und Geoökologie, Bereich Physische Geographie, Georgi-Dimitroff-Platz 1, Leipzig, DDR-7010.

[2]) Dr sc techn. NORBERT HANSEL, Technische Universität Dresden, Sektion Wasserwesen, Bereich Hydrologie und Meteorologie, Mommsenstraße 13, Dresden, DDR-8027

Резюме. Контроль ареалного загрязнения вод являемся серьёзной проблемой водого хозяйства и экологии человека. При этом загрязнение вод из-за сельского хозяйства особенно важно. Для контроля сельскохозяйственных источников загрязнения рекомендуются оборудование дренажных измери-тельных площадей в целях разных измерении и наблюдений дренажев или их частей.

Метод оборудования такой площади описан на примере дренажа Бёртевиц. Это дренажное поле, которые занимает площадь в 30 га и расположено в среднесаксонской лессовой области (210 до 220 м над уровнем моря).

Дренажные измерительные площади можно использовать

1 Как часть локальной специалной измерительной сети, напр. в территориях охраны питьевой воды,

2. в рамках обще государственной измерительной сети,

3. как агроэкологическая станция в целях исследования.

Названы критерия выбора, обрудования и работы дренажных измерительных площадей.

1. Das Problem

Die Kontrolle der flächenhaft bedingten Gewässerbelastung ist jetzt und in Zukunft ein ernsthaftes Problem, und zwar nicht nur für die Wasserwirtschaft, sondern auch in ökologischer Hinsicht. Dabei kommt der Belastung der Gewässer durch die Landwirtschaft eine wachsende Bedeutung zu, denn die Landwirtschaft muß zur besseren Ernährung der wachsenden Weltbevölkerung in allen Ländern der Erde intensiviert werden. Der atmosphärische Stoffeintrag in den Boden wächst z. Zt. sowohl nach der Masse als auch nach der Anzahl der eingetragenen Stoffe, die Schadstoffe inbegriffen. Es ist deshalb erforderlich:

- Grenzwertüberschreitungen durch Konzentrationsmessung sowie Frachtberechnung zu kontrollieren,
- hydrologische und stoffliche Bilanzen für den Oberboden aufzustellen,
- Entwicklungen und Trends der Belastungen zu analysieren,
- Ursachen der Belastungen und deren Veränderungen nachzuweisen und
- Maßnahmen zur Verringerung der flächenhaften Gewässerbelastung vorzuschlagen.

Eine praktikable Methode für solche Untersuchungen auf grundwasserfernen Standorten ist der Betrieb von Drainagemeßfeldern. Damit sind Messungen und Beobachtungen verschiedenen Umfangs an Drainagen oder Drainabteilungen gemeint.

2. Untersuchungsmethode

2.1. Drainagemethode

Eine direkte Messung des Sickerwasserabflusses erfolgt meist durch Lysimeter. Neben den hohen Baukosten solcher Anlagen ist eine Extrapolation der Meßergebnisse auf größere Flächen nur bedingt möglich. Demgegenüber bieten kleine hydrologische Einzugsgebiete zur Messung des Sickerwasserabflusses effektivere Ansätze. Dabei ist die hinreichende Abdichtung des geologischen Untergrundes wesentlichste Voraussetzung. Bei größeren Niederschlagsereignissen treten jedoch häufig Überlagerungen der Abflußkomponenten auf. Inwieweit dies stattfindet, wird durch bodengeologische und morphologische Bedingungen sowie die Größe des Einzugsgebietes bedingt. Drainierte lndwirtschaftliche Nutzflächen gestatten

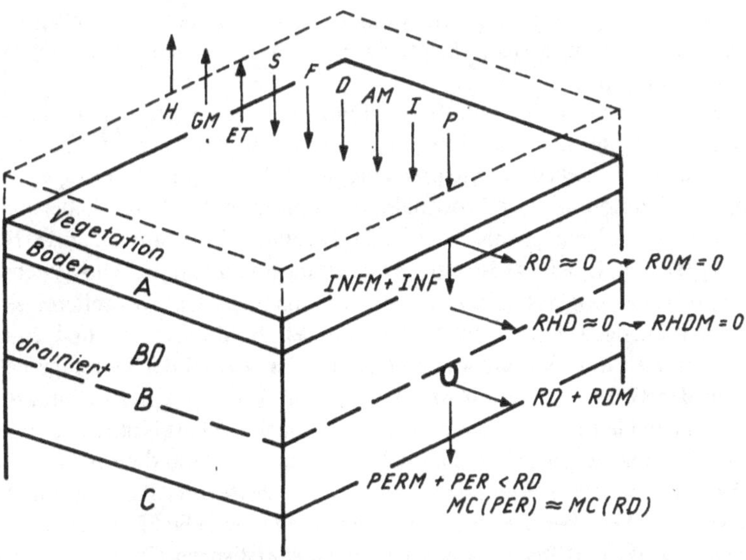

Abb. 1 Schematische Darstellung eines Drainagemeßfeldes

Input P – Niederschlag, I – Beregnung, D – organischer Dünger, F – mineralischer Dünger, S – Saatgut, AM – trockene atmosphärische Deposition

Output ET – Evapotranspiration, GM – gasförmiger Austrag, H – Ernte, RO – Oberflächenabfluß, ROM – Abspülung, INF + INFM – Infiltration und Einwaschung, RHD, RHIM – hypodermischer Abfluß und Auswaschung, RD + RDM – Drainageabfluß und Auswaschung, PER + PERM – Perkolation und Auswaschung, MC – Stoffkonzentration, A, BD, B, C – Bodenhorizonte

eine vom Oberflächenabfluß getrennte Messung des Sickerwasserabflusses. Hierzu sind folgende Bedingungen zu erfüllen (Abb. 1):

- Untersuchungsobjekt ist ein sehr kleines exakt abgrenzbares hydrologisches Einzugsgebiet,
- Einzugsgebietsgrenzen sind mit einem Drainagesystem oder Teilen davon identisch,
- Oberflächenabfluß und Erosion sind sehr selten und von untergeordneter Bedeutung (RO $\approx$ 0, ROM = 0),
- Bodendecke und Bewirtschaftung sind weitestgehend einheitlich,
- vertikale Abgrenzung besteht im Drainagesystem, starker Verdichtung des Unterbodens und des geologischen Untergrundes, so daß der Sickerwasserabfluß nahezu 0 ist (PER $\ll$ RD, PERM $\approx$ 0).

Untersuchungen nach diesen Bedingungen (GERDS [2]) gestatten an einem zentralen Meßpunkt, dem Drainageabfluß, alle abflußbildenden Faktoren, die innerhalb der Fläche variieren, integrativ zu erfassen. Eine ausreichende Abdichtung des Untergrundes liegt vor, wenn die hydraulische Leitfähigkeit (k_f) der unteren Schicht um mindestens 10^1 geringer als die der oberen ist (k_f(BD) > $10^1 \cdot$ $\cdot\, k_f$(B); HANSEL [3]).

Die vergleichsweise hohe Abflußvariabilität von Drainagen macht eine kontinuierliche Durchflußmessung notwendig. Die Beprobung kann in festgesetzten Zeitintervallen, aber auch zu entsprechenden Abflußereignissen erfolgen. Anzustreben ist eine automatisierte Probenahme mit programmierbarer Steuerung.

Ein Nachteil der Drainagemethode ist die Erfüllung der Bedingungen k_f(BD) > > $10^1 \cdot k_f$(B) und RO $\approx$ 0. Ist eine Abdichtung nicht ausreichend gegeben, so muß eine Abschätzung des „Fehlers" vorgenommen werden. Desweiteren können extreme Witterungsbedingungen zu Oberflächenabfluß, Erosion und somit zu nicht quantifizierbaren Stoffausträgen führen. Der Vorteil der Drainagemethode besteht in der Größe und somit der Aussagefähigkeit der Untersuchungsfläche. Im Vergleich zu kleinen Einzugsgebieten ist die Fläche seitens Nutzung und hydrologischen Regimes wesentlich einheitlicher und beinhaltet andererseits natürliche und ackerbauliche Inhomogenitäten. Es wird die reale Bewirtschaftung eines Feldes widergespiegelt. Darüber hinaus gestattet das Drainagemeßfeld die Untersuchung von Ernteprodukt und Bodenparametern nach statistischen Gesichtspunkten mit entsprechenden Stichprobenumfängen. Es liegt ein exakt definierbarer Bilanzraum vor.

2.2. Drainagemeßfeld Börtewitz

Vom Institut für Geographie und Geoökologie der AdW der DDR wurde eine Drainagemeßfläche ca. 60 km SE von Leipzig im Mittelsächsischen Lößhügelland ausgewäht. Die Fläche umfaßt mit 30 ha ein abgegrenztes Drainageeinzugsgebiet

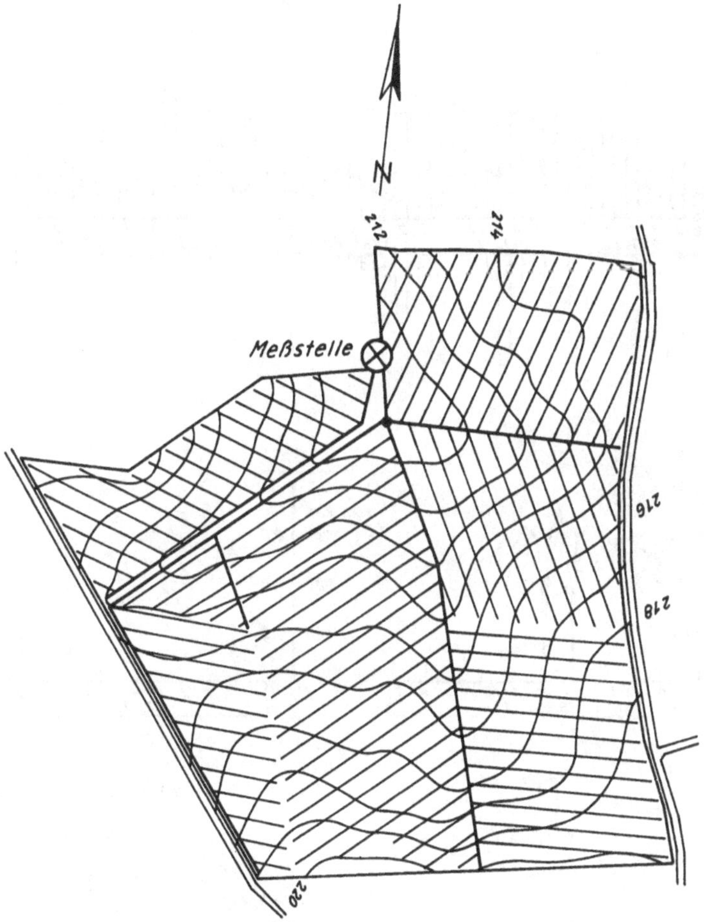

Abb 2. Drainagemeßfeld Börtewitz, Lageplan der Drainage

(Abb. 2). Die Bodendecke auf 4 bis 10 m mächtigem Lößlehm ist nur geringfügig differenziert. Ab 8 dm Tiefe treten Bodenverdichtungen ($\varrho > 1{,}6$ g/cm^3) auf. Das Drainagesystem ist in 1 m Tiefe mit 10 m Saugerabstand verlegt. Der geologische Untergrund ist kaolinisierter Quarzporphyr. Die Nutzung des Schlages besteht in einer Hackfrucht-Wintergetreide-Grünfutterrotation.

Der Drainageabfluß wird kontinuierlich durch ein Thompson-Wehr mit Trommelschreibpegel registriert. Die Beprobung erfolgt im 14tägigen Intervall und zu

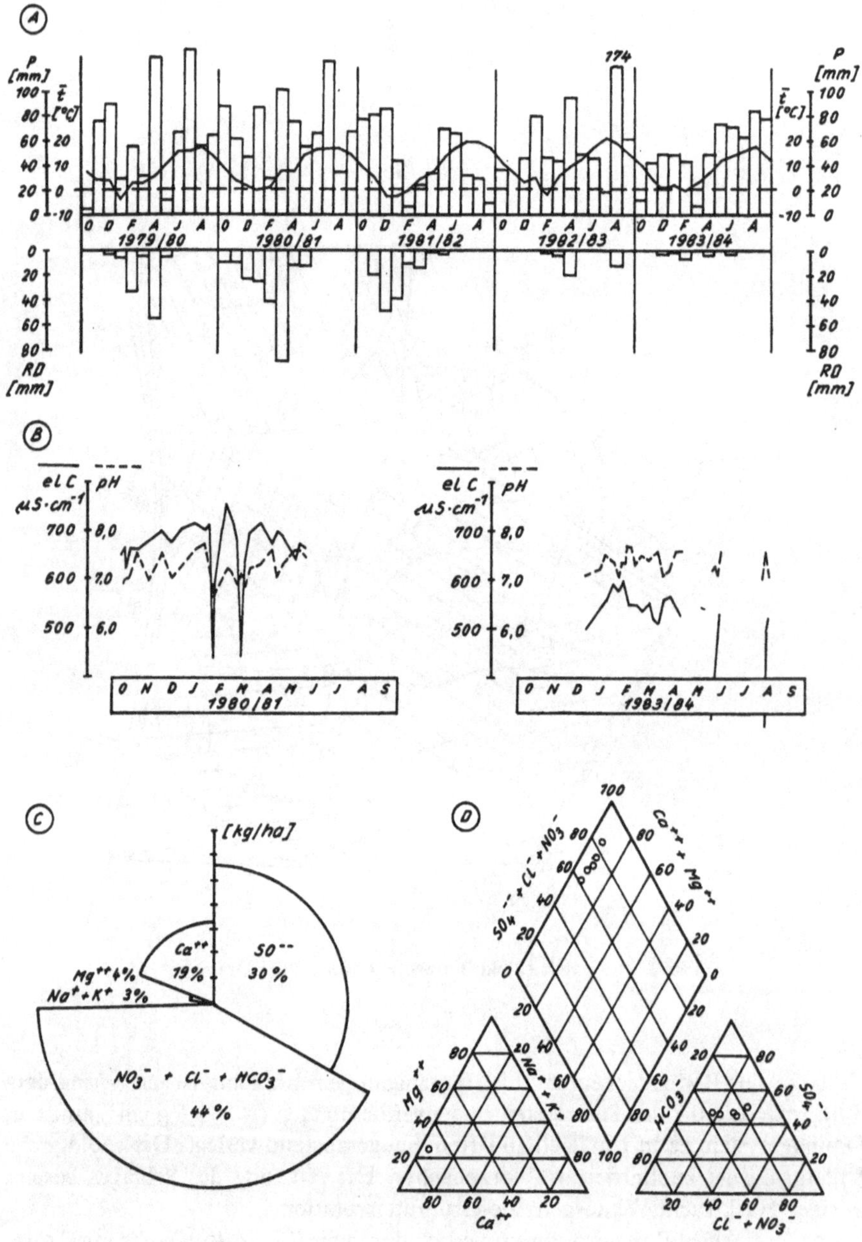
A
P [mm]
t̄ [°C]
RD [mm]
174
O D F A J A
1979/80
1980/81
1981/82
1982/83
1983/84
B
el C µS·cm⁻¹
pH
O N D J F M A M J J A S
1980/81
1983/84
C
[kg/ha]
Ca⁺⁺ 19%
SO⁻⁻ 30%
Mg⁺⁺ 4%
Na⁺+K⁺ 3%
NO₃⁻ + Cl⁻ + HCO₃⁻ 44%
D
SO₄⁻⁻ + Cl⁻ + NO₃⁻
Ca⁺⁺ + Mg⁺⁺
Mg⁺⁺
Na⁺ + K⁺
Ca⁺⁺
HCO₃⁻
SO₄⁻⁻
Cl⁻ + NO₃⁻

besonderen Abflußereignissen. Derzeit werden nur Makronährstoffe analysiert (Na, K, Ca, Mg, NH_4, NO_3, Cl, SO_4, HCO_3, PO_4). Die Niederschlagsmessung mit 2 bis 3 Sammelproben pro Monat erfolgt in unmittelbarer Nähe des Meßfeldes. Andere meteorologische Parameter werden von der nahegelegenen Station Oschatz des Meteorologischen Dienstes der DDR übernommen. Die Angaben über Stoffeintrag durch Düngung und Stoffaustrag durch Ernteprodukte stammen aus gesonderten Untersuchungen.

2.3. Untersuchungsergebnisse

Die Abflußmessungen zeigten, daß bis in Stundenintervalle eine sehr große zeitliche Variabilität besteht. Mehr als 3/4 der jährlichen Abflußmengen wurden in wenigen Tagen zwischen Dezember und April registriert (Abb. 3 A). Diese zeitliche Variabilität besteht jedoch auch zwischen den einzelnen Jahren. Durch die sehr hohe Speicherkapazität des Lößstandortes konnten in 5 Jahren Abflußmengen zwischen 220 und 20 mm gemessen werden. Das hat entscheidende Konsequenzen für die Menge der ausgewaschenen Nänrstoffe. Die zeitliche Veränderung der Konzentrationen ist im Vergleich zur Wassermenge sehr gering. Dies ist am Beispiel der elektrischen Leitfähigkeiten und der pH-Werte eines „Naß"- und eines „Trokkenjahres" zu sehen (Abb. 3 B). Daher ist die Zusammensetzung der jährlichen Frachten auch sehr einheitlich (Abb. 3 C). Bewirtschaftung und Witterung haben lediglich auf die einwertigen Anionen Einfluß. Dabei stellt sich je nach Bedingungen zwischen den NO_3- und HCO_3-Ionen ein Gleichgewicht ein. Dieser Effekt wird auch durch die hydrochemische Charakterisierung im Piper-Diagramm belegt (Abb. 3D). Durch die sehr hohe Sorptionskapazität des Lößbodens ist die Verteilung der Kationen innerhalb der untersuchten 5 Jahre fast gleich, während die der Anionen größeren Veränderungen unterliegt.

Über die hier angedeuteten Ergebnisse hinaus gestatten die vorliegenden Meßergebnisse die Aufstellung von Bilanzen aller analysierten Ionen. Dies ist insbesondere im Vergleich mit Untersuchungen von Bodennährstoffvorratsänderungen von Bedeutung.

Abb. 3. A — Monatssummen Niederschlag und Drainageabfluß sowie Monatsmittel der Lufttemperatur von Okt. 1979 bis Sept. 1984

B — Elektrische Leitfähigkeit und pH-Werte des Drainageabflusses eines feuchten (1980/81) und eines trockenen (1983/84) Jahres

C — Zusammensetzung der jährlichen Frachten des Drainabflusses (1979—1984)

D — Verteilung der Makroionen im Piper-Diagramm für fünf Untersuchungsjahre (1979—1984)

3. Anwendung zur Kontrolle der flächenhaften Gewässerverschmutzung

3.1. Prinzip der Anwendung

Die Anwendung von Drainagemeßfeldern zur Kontrolle der flächenhaften Gewässerverschmutzung entspricht der Beobachtung von Beispielsflächen. Das erscheint ökonomisch im Vergleich zu einer flächendeckenden Beprobung von Luft—Pflanze- Boden—Wasser nach den Regeln der statistischen Qualitätskontrolle.

Der Betrieb eines Drainagemeßfeldes hat eine Indikatorfunktion für den vergleichbaren Teil der landwirtschaftlichen Nutzfläche. Die allgemeine Anwendungsweise beschreibt Abb. 4.

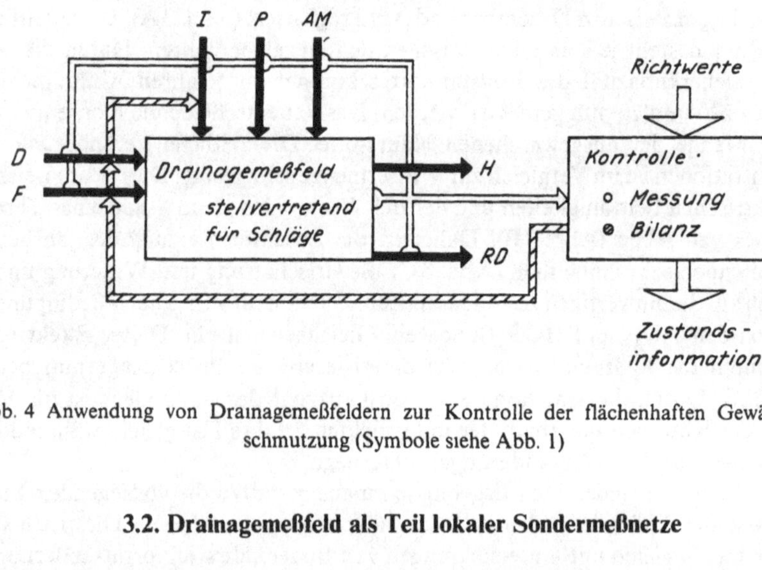

Abb. 4 Anwendung von Drainagemeßfeldern zur Kontrolle der flächenhaften Gewässerverschmutzung (Symbole siehe Abb. 1)

3.2. Drainagemeßfeld als Teil lokaler Sondermeßnetze

Drainagemeßfelder können als Teil von Sondermeßnetzen zur Kontrolle der flächenhaften Gewässerbelastung verwendet werden. Kontrolliert wird dabei das Drainagewasser in dem Moment, in dem es Oberflächenwasser wird. Solche lokalen Sondermeßnetze sind zweckmäßig in

- Einzugsgebieten von Trinkwasserfassungen,
- Heilquellenschutzgebieten,
- Schutzzonen von Trinkwassertalsperren,
- Kleinen Einzugsgebieten mit intensiver Mehrfachnutzung,
- Randgebieten industrieller und urbaner Ballungsräume oder auch einzelner Industriekomplexe.

3.3. Regionales Meßnetz aus Drainagemeßfeldern

Mit Drinagemeßfeldern kann ein regionales Meßnetz zur Kontrolle der flächenhaften Gewässerverschmutzung aufgebaut werden, das in der üblichen Weise hierarchisch strukturiert ist, indem es aus Stationen 1. bis 3. Ordnung besteht. Dabei können die Anzahl der Stationen und der Umfang der analysierten Stoffe örtlich und zeitlich veränderlich sein.

Das Meßnetz soll wenige Stationen 1. Ordnung haben, etwa eine für die wichtigsten Boden- bzw. Landschaftstypen des Landes. Die Drainagemeßfelder 1. Ordnung werden Modelldrainagen genannt, weil der hohe Meßaufwand die Modellierung der wichtigen Speicher-, Transport- und Reaktionsprozesse gestattet. Es erfolgen Messungen der Ein- und Ausgangsgröße sowie des Systemzustandes im erforderlichen Umfang mit hoher zeitlicher Auflösung. Damit sind möglich:

— Modelltestung und -parameterisierung,
— Zustandsbeschreibung der landwirtschaftlich aktiven Bodenzone und deren Änderung, auch infolge von Nutzungsänderungen,
— Entwicklung von Bilanzmethoden,
— Optimierung des Meß- und Beprobungsrhythmus (für 1. bis 3. Ordnung).

Zu dem Meßnetz sollen mehrere Stationen 2. Ordnung gehören, die nach geographischen Gesichtspunkten auszuwählen sind. Die Drainagemeßfelder 2. Ordnung werden Bilanzdrainagen genannt, weil ihr Beobachtungsumfang die statische Bilanz (Hansel [4]) von Wasser und von sickerwassergebundenen Stoffen für mittlere Zeitintervalle gestattet. Es erfolgen vorzugsweise Messungen der Ein- und Ausgangsgrößen in dem Umfang, der aus der geforderten Bilanzgenauigkeit folgt. Damit sind möglich:

— Tests von Bilanzmethoden,
— Aussagen zum innerjährlichen Gang der Bilanzgrößen,
— Mittelfristige Vorhersagen für die Landwirtschaft.

Stationen 3. Ordnung sollen aus Kostengründen im erforderlichen Mindestumfang in das Meßnetz aufgenommen werden. Sie sollen für die verschiedenen Standorte die typischen landwirtschaftlichen Produktionsbedingungen erfassen. Stationen 3. Ordnung werden Kontrolldrainagen genannt, da der Beobachtungsumfang quantifizierte Kontrollen gestattet. Es werden im wesentlichen Ausgangsgrößen gemessen. Damit sind möglich:

— Kontrollen der ausgetragenen Frachten und der Überschreitung von Grenzkonzentrationen,
— Aussagen anhand der Ganglinien von Mittelwerten mit größerem Zeitintervall.

Ein solches Meßnetz ist so aufgebaut, daß die Stationen niederer Ordnung den Meß- und Beobachtungsrhythmus der Stationen höherer Ordnung enthalten. Dadurch wird die Eichung der Stationen höherer Ordnung durch die Messungen an Stationen niederer Ordnung und die Regionalisierung der Ergebnisse möglich.

Ein solches Meßnetz muß längerfristig betrieben werden. Einerseits ist die Kontrolle der flächenhaften Gewässerverschmutzung auf lange Sicht eine wichtige Aufgabe. Andererseits erfordern die zeitliche Variabilität der meteorologischen und hydrologischen Größen, die Zeitdauer mehrerer aufeinanderfolgender Rotationen und die zu erwartenden Veränderungen eine vieljährige Beobachtung.

Schlußfolgernd kann die schrittweise Einrichtung eines Meßnetzes aus Drainagemeßfeldern als ständiges ökologisches Kontrollinstrument empfohlen werden. Der Ausbau kann dabei nach den regionalen Erfordernissen und den ökonomischen Möglichkeiten erfolgen. Dabei sollte in Schwerpunktgebieten (z. B. Trinkwasserschutzzonen) begonnen werden.

3.4. Drainagemeßfelder als agrarökologische Observatorien

Modelldrainagen können als agrarökologische Observatorien dienen und haben haben dabei als Forschungsinstrument eigenständige Bedeutung. Ihre Aufgaben sind dann:

- Kontrolle von Grenzwertüberschreitungen,
- Berechnung von Wasser- und Stoffbilanzen,
- Prozeßstudien unter Produktionsbedingungen,
- Modellentwicklung und -parametrisierung,
- Untersuchungen der Wechselwirkungen zwischen Luft—Pflanze—Boden—Wasser.

Drainagemeßfelder sind wiederholt als Forschungsinstrument genutzt worden (z. B. HEBERHOLD und KRAMER [5], ISTOK und KLING [6], MÜLLER et al. [7], MOLDAN und PACES [8]. SKAGGS et al. [9] und WORRESCHK[10]). Für agrarökologische Observatorien ist die Aufgabenstellung entsprechend zu erweitern.

Auf Zukunft hin betrachtet, sollten die ökologischen Observatorien der Agrarlandschaft durch ähnliche in der Wald-, Stadt- und Regenerationslandschaft ergänzt werden.

3.5. Gesichtspunkte für Auswahl, Einrichtung und Betrieb von Drainagemeßfeldern

Bei der Auswahl von Drainagemeßfeldern gelten folgende Gesichtspunkte:

- Die auszuwählende Fläche ist repräsentativ und definiert, seitens der Bodendecke möglichst einheitlich und übersichtlich.
- Die Drainage oder Dränabteilung ist langfristig funktionstüchtig. Neuanlagen sind zu bevorzugen.
- Bilanzfläche bzw. Bilanzraum sind weitgehend definiert.
- Die Morphologie ist dem Landschaftstyp entsprechend zu wählen, wobei $RO = RHD \approx 0$ erfüllt ist.

— Die Nutzung ist in der Fläche einheitlich und für regionale Produktionsbedingungen typisch.
— Entsprechend der Ordnung des Drainagefeldes ist es in der Nähe meteorologischer, hydrologischer und wasserwirtschaftlicher Stationen anzulegen. Die Datenerhebung soll kooperativ erfolgen.
— Die Beprobung ist jederzeit möglich; die An- und Abfahrt ist zeit- und energiesparend.
— Die Proben können mit vertretbarem Aufwand analysiert werden.

Bei der Einrichtung von Drainagemeßfeldern sollten nachstehende Hinweise berücksichtigt werden:
— Von Anfang an ist eine einheitliche Datengewinnung und -verdichtung sowie -konservierung aufzubauen.
— Jedes Drainagemeßfeld ist zu inventarisieren. Allgemeine Hinweise dazu können den DVWK-Regeln [1], speziellere solchen Vorschriften wie TGL 20286/01 und 02 [11] entnommen werden.
— Es sind Wasserproben zu entnehmen und nach TGL 28400/01 [12] zu analysieren sowie nach TGL 22764 [13] zu klassifizieren (oder nach analogen Normen).
— Der Drainageplan ist sorgfältig anzufertigen und dabei die Bilanzfläche sowie der Bilanzraum zu definieren. Dabei sind die Fließwege des Wassers zu betrachten für die Fälle: längere Trockenperiode — Abfluß aus ungesättigten Teilflächen — Abfluß aus gesättigten hypodermischen Teilflächen — Abfluß aus gesättigten Teilflächen — Grobporenabfluß — Sättigungsabfluß.
— Das Bodenprofil ist bis zu einer Tiefe von einigen Dezimetern unterhalb der Drainage aufzunehmen (etwa 1 Schurf pro 2 ha, mindestens 2 pro Drainagemeßfeld). Die Bodenproben sind physikalisch nach den Regeln der Pedologie zu charakterisieren. Zur bodenchemischen und -biologischen Kennzeichnung sind noch Festlegungen erforderlich.

Beim Betrieb von Drainagemeßfeldern sollte beachtet werden:
— Eine zentrale Anleitung ist erforderlich. Allgemeine Hinweise zur Auswertung u. a. können [1] entnommen werden.
— Der Meß- und Beprobungsrhythmus richtet sich nach der Ordnung des Drainagefeldes, der geforderten Genauigkeit und den zu analysierenden Stoffen. Gegebenenfalls ist der Aufwand während des Betriebes zu optimieren.
— Der Analyseumfang richtet sich nach dem atmosphärischen Stoffeintrag und dem Status der landwirtschaftlichen Produktion.
— Die Alterung der Drainagemeßfelder ist zu kontrollieren.

Literatur

[1] Empfehlungen zum Bau und Betrieb von Lysimetern. DVWK-Regeln zur Wasserwirtschaft 7/114. — Hamburg, Berlin: P. Parey, 1980.

[2] GERDS, W.: Zum sickerwassergebundenen Stoffumsatz von Agrarökosystemen. — Wiss. Mitt. IGG der AdW der DDR (Leipzig) (1984) 11.

[3] HANSEL, N.: Über natürliche Selbstdichtung von Gewässern mit porösem Untergrund und ihre hydrologisch-waaserwirtschaftliche Auswirkungen. — Wiss. Z. Hochsch. Bauwes. Leipzig (1967) 4, 29—41.

[4] HANSEL, N.: Möglichkeiten und Grenzen bei der Berücksichtigung von Ballungsräumen in der Modellierung des Wasserhaushaltes von Einzugsgebieten des Mittelgebirges. — Dresden: Technische Universität, Sektion Wasserwesen, Diss. B, 1981.

[5] HEBERHOLD, R.; KRAMER, U.: Auswirkungen unterschiedlicher Dränformen auf einen pseudovergleyten tonigen Boden. — Z. Kulturtechn. Flurber. (1983) 24.

[6] ISTOK, J. D.; KLING, G. F.: Effect of subsurface drainage on runoff and sediment yield from an agricultural watershed in Western Oregon, USA. — J. Hydrol. **65** (1983) 4.

[7] MÜLLER, G.; SCHULZE, F.; TRAEGER, M.: Bedeutung von Dränabflußmessungen im System der Grundwasserbeobachtung innerhalb landwirtschaftlich intensiv genutzter Einzugsgebiete. — Wasserwirtsch. Wassertechn. **32** (1982) 4.

[8] MOLDAN, B.; PACES, T.: Biogeochemischer Stoffkreislauf in einem landwirtschaftlich genutzten Beispielgebiet. — Praha: Ustredni Ustav Geologicky, 1980.

[9] SKAGGS, R. W.; et al.: Effect of drainage system design on surface and subsurface runoff from artificially drained lands. — In: Proc. of the Int. Symp. on Rainfall Runoff Modelling. — Mississippi, USA: Water Res. Publ., 1982.

[10] WORRESCHK, B.: Beitrag zur Berechnung der Abflußbildung in landwirtschaftlich genutzten Einzugsgebieten. — Mitt. Leichtweiss Inst. f. Wasserbau der TU Braunschweig (1985) 85.

[11] Fachbereichsstandard der DDR: Melioration, Drainagen. Grundsätze für die Projektierung und Bemessung der Rohrdränung. — TGL 20286/01 und 20286/02, 1978.

[12] DDR-Standard: Wasseruntersuchungen — Grundsätze. — TGL 28400/01, 1973.

[13] Fachbereichsstandard der DDR: Nutzung und Schutz der Gewässer. Klassifizierung der Wasserbeschaffenheit von Fließgewässern. — TGL 22764, 1981.

Acta hydrophys., Berlin **31** (1987) 3/4, S. 251—254

Eine Bemerkung zur modernen Seegangsforschung

Von H. GIENAPP[1])

In einer früheren Arbeit (GIENAPP [1]) hat der Verfasser Ergebnisse der Berechnungen von Amplitudenspektren von natürlichen Wasserwellen beschrieben, die bei einem schweren Sturm mit Windstärken bis zu Beaufort 10 in flachem Wasser (10 m tief) der Deutschen Bucht bei der Außeneider am Vormittag des 2. November 1965 gemessen worden sind. Die Wetterlage ist in Abb. 1 und Tabelle 1 dargestellt. Sie war über der Deutschen Bucht durch ein sehr homogenes Windfeld von westlicher Richtung gekennzeichnet.

Charakteristisch für die analysierten Amplitudenspektren von 2 Minuten langen Registrierungen des Seegangs war, daß sich sämtliche auftretenden Frequenzen als Ober- bzw. Kombinationsschwingungen von 2 Grundschwingungen darstellen ließen, nämlich einer Dünungs- und einer Windseeperiode, bis auf eine zusätzlich auftretende Welle mit einer Schwingungszeit von 5—6 s. Die Realität dieses Effektes ist lange mit wechselnden Argumenten bestritten worden (H. WALDEN, K. RICHTER, H. CARLSON usw.), obgleich Fehler am Meßgerät oder an der Rechenmethode nicht nachgewiesen werden konnten. Als Ursache des Phänomens sind damals vom Verfasser starke Wechselwirkungen im Anregungsprozeß von Wasserwellen durch Wind vermutet worden.

In der Zwischenzeit seit dem Erscheinen dieser Arbeit hat sich nun ein neues Gebiet der Physik entwickelt, das geeignet ist, die Erscheinungen zwischen dem (Seegangs-)Chaos und einer determinierten Einzelwelle zu beschreiben, nämlich die Synergetik (HAKEN [3], [4]; SCHUSTER [5]; GROSSMANN [2]).

Ein typisches synergetisches Phänomen ist das Auftreten subharmonischer Wellen mit verdoppelten, verdreifachten Perioden usw. aufgrund nichtlinearer Prozesse als Vorboten des Chaos sowie der Abbau dieser (Periodenverdoppelungs-) Kaskade bei noch weitergehender Annäherung an das Chaos.

---

[1]) Dr. rer. nat. HANS GIENAPP, Deutsches Hydrographisches Institut Hamburg, Postfach 220, D-2000 Hamburg 4.

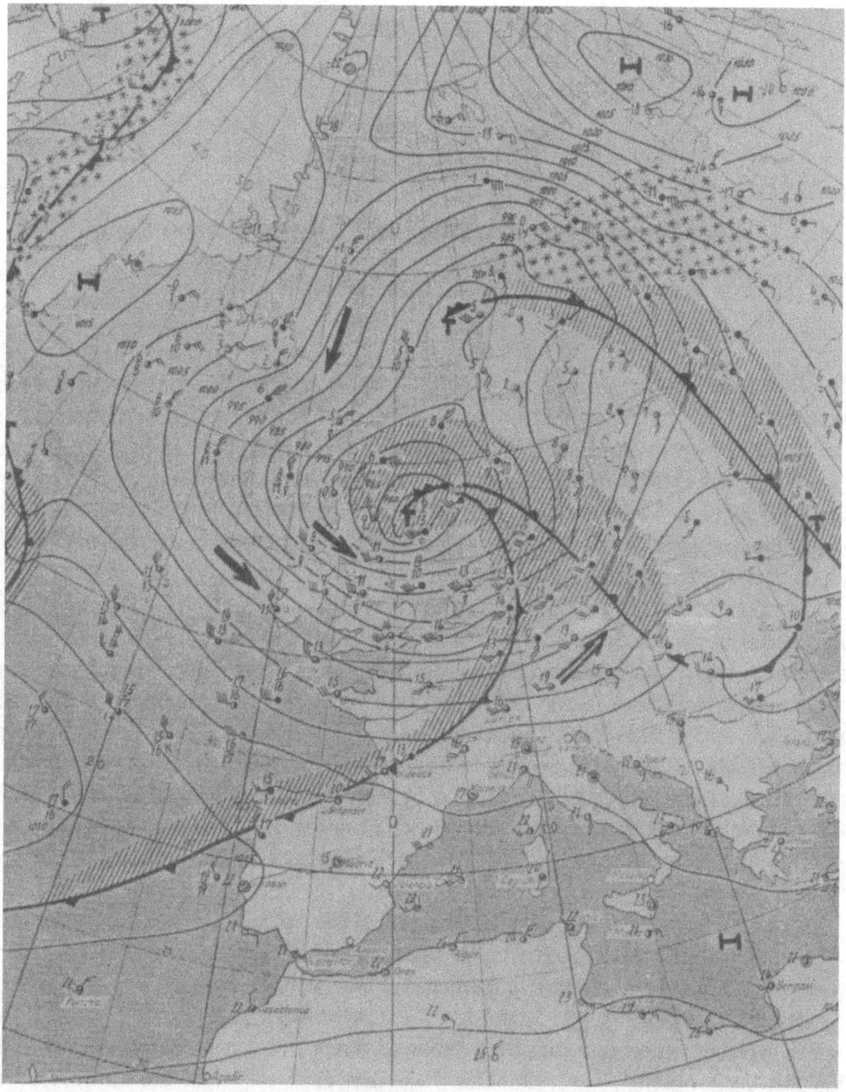

Abb. 1. Wetterlage über der Nordsee am 2. November 1965, 13 Uhr

Dieser Umstand führt auf den Gedanken, die dritte Grundwelle in den beschriebenen Amplitudenspektren als Kombinationswelle der ersten subharmonischen der Seegangswelle mit der Dünung aufzufassen, s. Tabelle 2. Die Übereinstimmung

TABELLE 1
Wetterablauf am Vormittag des 2. November 1965

Tag	Montag, 1 11 1965				Dienstag, den 2. November 1965															
Uhrzeit	19 Uhr				1 Uhr				7 Uhr				13 Uhr				13 Uhr			
																		Wellen-	
Feuer-schiffe	Wind u. Temperatur (zum Termin), Wetterablauf (während der letzten 6 Stunden)																Sicht km	Per sec	Höhe m
Borkumriff	W	10	12	Schauer	W	10	10	Schauer	W	9	10	Schauer	NW	6	9	Schauer	18	6	4,5
Weser	W	8	12	Schauer	W	9	9	Regen	W	9	9	Regen							
P8 (54,3 N 7,2 E)	WSW	10	11	Regen	W	10	10	Regen											
P12 (54,0 N 7,9 E)	SW	10	13	Schauer	W	10	10	Schauer	W	9	10	Schauer	WNW	6	8	Schauer	10	6	3
Elbe 1	W	8	13	bewölkt	WSW	10	11	Schauer	W	9	10	Regen	NW	9	10	Schauer	10	4	0,2
Kiel	SW	9	12	Regen	WSW	8	10	Schauer	WSW	8	9	Schauer	WSW	6	9	Regen	18	4	1
Fehmarnbelt	SW	8	12	Schauer	WSW	9	11	w, bew	WSW	9	9	Schauer	W	7	10	Schauer	15	4	1,5

Tabelle 2
Die Darstellung der dritten Welle als Additionswelle von Dünung und subharmonischer Seegangswelle

	Messung Nr.				
	334	335	336	337	338
Dünungsfrequenz*)	13	13	11	11	12
Halbe Seeg.-Frequenz	8,5	9,5	8	7	8
Summe Zeile 1 und Zeile 2	21,5	22,5	19	18	20
Meßwert der Frequenz 3. Welle	22	24	18	20	20
Differenz Zeile 5—Zeile 4	0,5	1,5	−1	2	0

*) Gemessen in relativen Einheiten, einer Frequenz 12 entspricht 10,0 s Schwingungszeit

der analysierten mit den auf diese Weise kombinierten Frequenzen ist sehr gut. Die Summe der Fehlerquadrate ist $0{,}5^2 + 1{,}5^2 + 1^2 + 2^2 + 0^2 = 7{,}5$. Berechnet man auf entsprechende Weise zur Kontrolle die Fehlerquadratsumme für den Fall der Kombination der Seegangsfrequenz mit der halben Dünungsfrequenz, so ergibt sich der wesentlich größere Wert 21,0. Das ist sehr plausibel, weil die Dünungswelle in viel höherem Maße linear ist als die Seegangswelle, also nur in geringerem Maße zu synergetischen Effekten Anlaß geben kann.

Der Verfasser hofft, mit dieser Betrachtung zur Neubelebung der Seegangsforschung beizutragen.

Literatur

[1] Gienapp, H.: Flachwasserseegang bei Sturm. — Z. angew. Physik **28** (1969), 117—123.

[2] Grossmann, S.: Chaos — Unordnung und Ordnung in nichtlinearen Systemen. — Phys. Bl. **39** (1983), 139—145.

[3] Haken, H.: Synergetik — Eine Einführung. — Berlin, Heidelberg, New York: 1982.

[4] Haken, H.: Evolution of Order and Chaos. — Berlin, Heidelberg, New York: Proc. Int. Symp. on Synergetics, 1982.

[5] Schuster, H. G.: Deterministic Chaos. — Weinheim: 1984.

FSC
www.fsc.org
MIX
Papier | Fördert
gute Waldnutzung
FSC® C083411

Zeitfracht Medien GmbH
Ferdinand-Jühlke-Straße 7
99095 Erfurt, Deutschland
produktsicherheit@kolibri360.de